Hands-on Question Answering Systems with BERT

Applications in Neural Networks and Natural Language Processing

Navin Sabharwal
Amit Agrawal

Apress®

Hands-on Question Answering Systems with BERT

Navin Sabharwal
New Delhi, Delhi, India

Amit Agrawal
Mathura, India

ISBN-13 (pbk): 978-1-4842-6663-2   ISBN-13 (electronic): 978-1-4842-6664-9
https://doi.org/10.1007/978-1-4842-6664-9

Managing Director, Apress Media LLC: Welmoed Spahr
Acquisitions Editor: Celestin Suresh John
Development Editor: Matthew Moodie
Coordinating Editor: Aditee Mirashi

Cover designed by eStudioCalamar

Cover image designed by Freepik (www.freepik.com)

Distributed to the book trade worldwide by Springer Science+Business Media New York, 1 New York Plaza, Suite 4600, New York, NY 10004-1562, USA. Phone 1-800-SPRINGER, fax (201) 348-4505, e-mail orders-ny@springer-sbm.com, or visit www.springeronline.com. Apress Media, LLC is a California LLC and the sole member (owner) is Springer Science + Business Media Finance Inc (SSBM Finance Inc). SSBM Finance Inc is a **Delaware** corporation.

For information on translations, please e-mail booktranslations@springernature.com; for reprint, paperback, or audio rights, please e-mail bookpermissions@springernature.com.

Apress titles may be purchased in bulk for academic, corporate, or promotional use. eBook versions and licenses are also available for most titles. For more information, reference our Print and eBook Bulk Sales web page at http://www.apress.com/bulk-sales.

Any source code or other supplementary material referenced by the author in this book is available to readers on GitHub via the book's product page, located at www.apress.com/ 978-1-4842-6663-2. For more detailed information, please visit http://www.apress.com/ source-code.

Printed on acid-free paper

Dedicated to the people I love and the God I trust.
—Navin Sabharwal

Dedicated to my family and friends.
—Amit Agrawal

Table of Contents

About the Authors

Navin Sabharwal is the Chief Architect for HCL DryICE Autonomics. He is an innovator, thought leader, author, and consultant in the areas of artificial intelligence (AI), machine learning, cloud computing, big data analytics, and software product development. He is responsible for Intellectual Property (IP) development and service delivery in the areas of AI and machine learning, automation, AIOps, public cloud Google Cloud Platform (GCP), Amazon Web Services (AWS), and Microsoft Azure. Navin has authored more than 15 books in the areas of cloud computing, cognitive virtual agents, IBM Watson, GCP, containers, and microservices. He is reachable at Navinsabharwal@gmail.com.

Amit Agrawal is a principal data scientist and researcher delivering solutions in the fields of AI and machine learning. He is responsible for designing end-to-end solutions and architecture for enterprise products. He can be reached at agrawal.amit24@gmail.com.

About the Technical Reviewer

Riya Naval is a Senior Data Scientist practicing in the latest AI technologies. She is responsible for designing, developing, and delivering an end-to-end solution based on AI. She can be contacted at **https://www.linkedin.com/in/ riya-naval-010ab2b4**.

Acknowledgments

I would like to thank my family, Shweta and Soumil, for always being by my side, for sacrificing their time for my intellectual and spiritual pursuits, and for taking care of everything while I was immersed in authoring this book. This and other accomplishments in my life wouldn't have been possible without their love and support. I also thank my mom and my sister for their love and support, as always; without their blessings, nothing is possible.

To my coauthor, Amit, I thank you for the hard work and quick turnarounds to deliver this book. It was an enriching experience, and I look forward to working with you again soon.

Thanks go to all of my team members, who have been a source of inspiration with their hard work, their always engaging technical conversations, and their technical depth. Your constantly flowing ideas are a source of encouragement and excitement every single day. Piyush Pandey, Sarvesh Pandey, Amit Agrawal, Vasand Kumar, Punith Krishnamurthy, Sandeep Sharma, Amit Dwivedi, Gauarv Bhardwaj, Nitin Narotra, Divjot, and Vivek, thank you for being there and making technology fun.

To all my other coauthors, colleagues, managers, mentors, and guides, in this world of 7 billion, it was coincidence that brought us together, but it has been an enriching experience to be associated with you and learn from you. All ideas and paths are an assimilation of conversations that I have had and experiences I have shared. Thank you.

—*Navin Sabharwal*

ACKNOWLEDGMENTS

I thank my parents, brothers, and wife Riya for always being an inspiration for me.

Thanks go to my coauthor, Navin, for his guidance and feedback.

I am also grateful to my colleagues Rishabh Upadhyay and Yogita Kanwar for their technical suggestions.

—*Amit Agrawal*

Thank you goddess Saraswati, for guiding me to the path of knowledge and spirituality: असतो मा साद गमय, तमसो मा ज्योतिर् गमय, मृत्योर मा अमृतम् गमय

Asato Ma Sad Gamaya, Tamaso Ma Jyotir Gamaya,

Mrityor Ma Amritam Gamaya

Lead us from ignorance to truth, lead us from

darkness to light, lead us from illusion to reality

Introduction

Question answering systems have revolutionized information retrieval. Technologies like Bidirectional Encoder Representations from Transformers (BERT) have made it possible for documents to be analyzed by machine learning systems and retrieve contextual information through the question-and-answer mechanism without the need for extensive training. Evolution of deep learning has had a great impact on the design of question answering systems and has enabled these systems to ingest enormous amounts of data and build billions of connections to understand human language better.

This book focuses on a recent breakthrough in the natural language processing (NLP) domain, BERT which has achieved benchmarks on state-of-art NLP tasks such as question answering system, entity recognition systems, and so on.

BERT implements innovative ways to generate the embedding of textual sentences. This book provides guidance on design and implementation of various types of question answering systems along with NLP tasks such as document summarization, entity recognition, and sentiment analysis. This could help data scientists and developers to design and implement their own NLP-based systems using BERT.

Let's start our journey in the exciting and quickly evolving domain of NLP.

CHAPTER 1

Introduction to Natural Language Processing

With recent advances in technology, communication is one of the domains that has seen revolutionary developments. Communication and information have formed the backbone of modern society and it is language and communication that has led to such advances in human knowledge in all spheres. Humans have been fascinated by the idea of machines or robots having human-like abilities to converse in our language. Numerous science fiction books and media have dealt with this topic. The Turing test was designed for this purpose, to test whether a human being is able to decipher if the entity on the other end of a communication channel is a human being or a machine.

With computers, we started with a binary language that a computer could interpret and then compute based on the instructions. Over time, however, we came up with procedural and object-oriented languages that use syntax and instructions in languages that are more natural and correspond to the words and ways in which humans communicate. Examples of such constructs are for loops and if constructs.

© Navin Sabharwal, Amit Agrawal 2021
N. Sabharwal and A. Agrawal, *Hands-on Question Answering Systems with BERT*,
https://doi.org/10.1007/978-1-4842-6664-9_1

With the availability of increased computing capacity and the ability of computers to process huge amounts of data, it became easier to use machine learning (ML) and deep learning models to understand human language. With neural networks, recurrent neural networks (RNNs), and other deep learning technologies becoming accessible and the computing power to run these models available, a variety of natural language processing (NLP) platforms became available for developers to work with over the cloud and on premises. This chapter takes you through the basics of NLP.

Natural Language Processing

NLP is a sub-branch of artificial intelligence (AI) that enables computers to read, understand, and process human language. It is very easy for computers to read data from structured systems such as spreadsheets, databases, JavaScript Object Notation (JSON) files, and so on. However, a lot of information is represented as unstructured data, which can be quite challenging for computers to understand and generate knowledge or information. To solve these problems, NLP provides a set of techniques or methodologies to read, process, and understand human language and generate knowledge from it. Currently, numerous companies including IBM, Google, Microsoft, Facebook, OpenAI, and others have been providing various NLP techniques as a service. Some open-source libraries such as NLTK, spaCy, and so on are also key enablers in making it possible to break down and understand the meaning behind linguistic texts.

As we know, processing and understanding of text is a very complex problem. Data scientists, researchers, and developers have been solving NLP problems by building a pipeline: breaking up an NLP problem into smaller parts; solving each of the subparts with their corresponding NLP techniques and ML methods such as entity recognition, document summarization, and so on; and finally combining or stacking all parts or models together as the final solution to the problem.

The main objective of NLP is to teach machines how to interpret and understand language. Any language such as English, programming construct, mathematics, and so on, involves the following three major components:

- **Syntax**: Defines rules for ordering of words in text. As an example, subject, verb, and object should be in the correct order for a sentence to be syntactically correct.

- **Semantics**: Defines the meaning of words in text and how these words should be combined together. As an example, in the sentence, "I want to deposit money in this bank account," the word "bank" refers to a financial institution.

- **Pragmatics**: Defines usage or selection of words in a particular context. As an example, the word "bank" can have different meanings on the basis of context. For example, "bank" could also mean financial institution or land at the edge of a river.

For this reason, NLP employs different methodologies to extract these components out of text or speech to generate features that will be used for downstream tasks such as text classification, entity extraction, language translation, and document summarization. Natural language understanding (NLU), a sub-branch of NLP that aims at understanding and generating knowledge from documents, web pages, and so. Some examples are listed here.

- **Language translation**: Language translation is considered one of the most complex problems in NLP and NLU. You can provide text snippets or documents and these systems will convert them into another language. Some of the major cloud vendors such as

Google, Microsoft, and IBM provide this feature as a service that can be leveraged by anyone for their NLP-based system. As an example, a developer who is working on development of a conversation system can leverage translation services from these vendors to enable multilingual capability in a conversation system without even doing any actual development.

- **Question-answering system**: This type of system is very useful if you want to implement a system to find an answer to a question from a document, paragraph, database, or any other system. Here, NLU is responsible for understanding a user's query as well as the document or paragraph (unstructured text) that contains the answer to that question. There exist a few variations of question-answering systems, such as reading comprehension-based systems, mathematical systems, multiple choice systems, question-answering and so on.

- **Automatic routing of support tickets**: These systems read through the contents of customer support tickets and route them to the person who can solve the issue. Here, NLU enables these systems to process and understand emails, topics, chat data, and more, and route them to the appropriate support person, thereby avoiding extra hops due to incorrect assignation.

Systems such as question-answering systems, machine translation, named entity recognition (NER), document summarization, parts of speech (POS) tagging, and search engines are some of examples of NLP-based systems.

As an example, consider the following text from the Wikipedia article for "Machine Learning".

> Machine learning (ML) is the scientific study of algorithms and statistical models that computer systems use to perform a specific task without using explicit instructions, relying on patterns and inference instead. Machine learning algorithms are used in a wide variety of applications, such as email filtering and computer vision. It can be divided into two types, i.e., Supervised and Unsupervised Learning.

This text includes a lot of useful data that can be used as information. It would be good if computers could read, understand, and answer the following questions from the text:

- What are the applications of machine learning?
- What type of study does machine learning refer to?
- What type of models do computers use to perform specific tasks?

There should be some way to teach a machine the basic concepts and rules of languages so that they can read, process, and understand text. To derive an insight from a text, NLP techniques combine all of the steps into a pipeline known as the NLP/ML pipeline. The following are some of the steps of an NLP pipeline.

- Sentence segmentation
- Tokenization
- POS tagging
- Stemming and lemmatization
- Identification of stop words

Sentence Segmentation

The first step in the pipeline is to segment the text snippet into individual sentences, as shown here.

- Machine learning (ML) is the scientific study of algorithms and statistical models that computer systems use to perform a specific task without using explicit instructions, relying on patterns and inference instead.

- Machine learning algorithms are used in a wide variety of applications, such as email filtering and computer vision.

- It can be divided into two types, i.e., Supervised and Unsupervised Learning.

Earlier implementation of sentence segmentation was quite easy, just splitting the text on the basis of punctuation, or a "full stop." Sometimes that failed, though, when documents or a piece of text were not formatted correctly or were grammatically incorrect. Now, there are some advanced NLP methods such as sequence learning that segments a piece of text even if a full stop is not present or a document is not formatted correctly, basically extracting phrases by breaking up text using semantic understanding along with syntactic understanding.

Tokenization

The next task in the NLP pipeline is tokenization. In this task, we break each of the sentences into multiple tokens. A token can be a character, a word, or a phrase. The basic methodology used in tokenization is to split a sentence into separate words whenever there is a space between them. As an example, consider the second sentence from our example text:

"Machine learning algorithms are used in a wide variety of applications, such as email filtering and computer vision." Here is the result of applying tokenization to this example.

```
["Machine", "learning", "algorithms", "are", "used", "in" ,
"a", "wide", "variety", "of", "applications", "such", "as",
"email", "filtering", "and", "computer", "vision"].
```

However, there are some advanced tokenization methods such as Markov chain models that can extract phrases out of a sentence. As an example, "machine learning" can be extracted as a phrase by applying advanced ML and NLP methods.

Parts of Speech Tagging

POS tagging is the next step to determine parts of speech for each of the tokens or words extracted from the tokenization step. This helps us to identify the use of each word and its significance in a sentence. It also introduces first steps toward the actual understanding of the meaning of a sentence. Imparting a POS tag can increase the dimension of the word, to give better detail of the meaning the given word is trying to impart. The phrases "putting on an act" and "act on an instinct" both use the word "act," but as a noun and a verb, respectively, so a POS tag can greatly help in distinguishing the meaning. In this approach, we pass the token, referred as Word, to the POS tagger, a classification system, along with some context words that will be used to classify the Word with its relevant tags as shown in Figure 1-1.

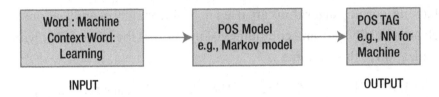

Figure 1-1. *POS tagging*

These models are trained on a huge corpus of (millions or billions) sentences of literature in the target language where each word along with its POS tag is used as training data for the POS classifier. The previously mentioned models are completely based on statistics as per training data and not on actual interpretation. The model tries to find POS tag for each of the words based on syntactic similarity of a sentence with historical sentences. As an example, for the sentence "Machine learning algorithms are used in a wide variety of applications, such as email filtering and computer vision," the POS tag is as shown here:

```
Machine (NN) learning (NN) algorithms (NNS) are (VBP)
used (VBN) in (IN) a (DT) wide (JJ) variety (NN) of (IN)
applications (NNS), such (JJ) as (IN) email (NN) filtering
(VBG) and (CC) computer (NN) vision (NN).
```

As we can see from those results, there are various nouns (i.e., *Machine, learning, variety, computer,* and *vision*). We can thus conclude that the sentence may be talking about machines and computers.

Stemming and Lemmatization

Sometimes the same word occurs in multiple sentences in different forms. Stemming can be defined as the process of reducing words to their root or base form by removing suffixes. Here, the reduced words can be dictionary words or nondictionary words. For example, the word "machine" can be reduced to its root form, "machin." It doesn't take into consideration the context in which word is being used. Here is the stemmed representation of tokenized words for our example sentence.

```
machin learn algorithm ar us in a wid vary of apply , such as
email filt and comput vis
```

In this result, some of the words are represented as nondictionary words; for example, "machine" reduced to "machin," which is a stemmed word but not a dictionary word.

Lemmatization can be defined as a process of deriving a canonical form or lemma of the word. It uses context to identify the lemma of the word, which must be a dictionary word. However, the same is not true for stemming. Using our previous example, the word "machine" will be converted into its canonical form as "machine." The following is the lemmatized representation of tokenized words of our example sentence. It uses tags of words as context to derive canonical forms of words.

```
Machine learning algorithm be use in a wide variety of
application , such a email filtering and computer vision.
```

In these results, some of the words, such as "filtering," are reduced to their canonical form, in this case "filtering," not "filter," because the word "filtering" is being used as a verb in the sentence.

Lemmatization and stemming should be used with utmost care and as per requirements. For example, if you are working with a search engine system, then stemming should be preferred, but if you are working with question answering, where reasoning is important, then lemmatization should be preferred over stemming.

Identification of Stop Words

Text snippets contain important as well as filler words. For example, in our example sentence, these are the filler words.

```
["be", "use", "in", "a", "such", "a", "and"]
```

These filler words introduce noise into your text and it is important to manage them, as they appear very frequently in the text and will have a much higher frequency and less importance than other words. Some systems use a predefined list of these stop words, such as "is," "at," and so on. This is not helpful for some domains, though. As an example, in documents related to health care, you will find some common terms such as patient, doctor, or ICU. These words appear very frequently and you need to somehow remove them from your text. There are two methods that are generally used to deal with domain-specific stop words.

- Flag words as stop words on basis of their frequency of occurrence. It could be either most frequent or least frequent.

- Flag words as stop words if they are quite common across all documents in the corpus.

Phrase Extraction

Sometimes a single word doesn't provide sufficient information for most of the NLP tasks. As an example, the meaning of the two words "machine" and "learning" from the dictionary are shown here.

- **Machine**: An apparatus using mechanical power to perform certain tasks.

- **Learning**: An acquisition of knowledge or skills through study, experience, or being taught.

It is very clear from the definitions of these two words that our sample sentence should have been talking about some mechanical device and various media for acquiring the knowledge. However, when these words are used together (i.e., "machine learning"), it refers to the sub-branch of AI that deals with the scientific study of algorithms and statistical models used by computers to perform a specific task without being explicitly programmed.

To extract phrases, we need to combine multiple words together, or identify phrases. Here, phrases can be of two types, noun phrases and verb phrases. We can define rules to extract phrases from sentences. As an example, to extract a noun phrase, we can define a rule such that "two consecutive occurrences of nouns in a sentence should be considered a noun phrase." For example, the phrase "machine learning" is a noun phrase in our sample sentence. In a similar manner, we can define more rules to extract noun phrases and verb phrases from a sentence.

Named Entity Recognition

An entity is defined as an object or noun such as a person, organization, or other object that provides important information from the text. This information can be used as a feature for downstream tasks. As an example, Google, Microsoft, and IBM are entities of the type *Organization*.

NER is an information extraction technique that extracts and classifies entities into categories as per the trained model. As an example, some of the basic categories in the English language are names of persons, organizations, locations, dates, email addresses, phone numbers, and so on. For example, in our sample sentence phrases such as "machine learning" and "computer vision" are entities of type *AI_Branch,* which refers to branches of AI.

Currently, large vendors in the AI domain such as IBM, Google, and Microsoft provide their trained models to extract named entities from the text. They also enable you to build your own NER model specific to your application and domain. Open-source projects such as spaCy also provide the capability to train and use your own custom NER model.

Coreference Resolution

One of the major challenges in the NLP domain, especially in the English language, is the use of pronouns. In English, pronouns are used extensively to refer to nouns in a previous context or sentence. To perform semantic analysis or identify the relationship between these sentences, it is very important that somehow the system should establish dependencies between the sentences.

As an example, consider the sentence "It can be divided into two types, i.e., Supervised and Unsupervised Learning," where "It" refers to machine learning in the first and second sentences. It can be accomplished by annotating such dependencies in the dataset for training a model and using the same model over unseen text snippets or documents to extract such relationships.

Bag of Words

As we all know, computers work on numerical data only; therefore, to understand meaning of text, it must be converted into a numerical form. Bag of words is one of the approaches for converting text into numerical data.

Bag of words is a very popular feature extraction method that describes the occurrence of each word in the text. You need to first build the vocabulary of your corpus then calculate the occurrence of each word corresponding to each text snippet or document in the corpus. It doesn't store any information related to order or sentence structure. That's why it is known as a bag of words. It can also tell you whether a particular word is present in the document or not, but it doesn't provide any information about the location of the word in the document. As an example, consider our example text snippet, which has been segmented into three sentences as a result of the sentence segmentation step.

- **Sentence A**: Machine learning (ML) is the scientific study of algorithms and statistical models that computer systems use to perform a specific task without using explicit instructions, relying on patterns and inference instead.

- **Sentence B**: Machine learning algorithms are used in a wide variety of applications, such as email filtering and computer vision.

- **Sentence C**: It can be divided into two types, i.e., Supervised and Unsupervised Learning.

Figure 1-2 is a document-term matrix for our example text snippet, where the term value is 1 if it is present in the sentence, or 0 otherwise.

	models	ml	patterns	perform	divided	algorithms	...	scientific	systems
Sentence A	1	1	1	1	0	1	...	1	1
Sentence B	0	0	0	0	0	1	...	0	0
Sentence C	0	0	0	0	1	0	...	0	0

Figure 1-2. *Document-term matrix*

Once sentences or text snippets are converted into vectors of numbers, we can use these vector values as a feature for further downstream tasks such as a question-answering system, text summarization, and so on. This method has the following limitations.

- Length of vector representation for the sentence increases as vocabulary size increases. This requires higher computation for downstream tasks. It also increases dimensionality of sentences.

- It can't identify different words with similar meanings on the basis of their context in the text.

There are other methods that reduce computation and memory requirements to represent sentences in vector form. Word embedding is one of the approaches where we can represent a word in lower dimensional space while preserving the semantic meaning of the word. We will see in detail later how word embedding is major breakthrough for downstream NLP tasks.

Conclusion

This chapter discussed the basics of NLP, along with some of the basic NLP tasks such as tokenization, stemming, and more. In next chapter, we discuss neural networks in the NLP domain.

CHAPTER 2

Neural Networks for Natural Language Processing

Bringing human cognitive intelligence (i.e., thinking, reasoning, and action) to artificial systems has always been a hot topic for researchers. In this process, they came up with idea of neural networks that try to emulate how the neurons of the human brain work. Although they are still very far from human cognitive capability, artificial neural networks hold a very promising position in the area of ML, and have transformed the way NLP applications are developed.

In this chapter, we will discuss neural networks and their types, along with some special types of neural networks, such as long short-term memory (LSTM), convolutional neural networks (CNNs), encoders, decoders, and transformers. This will set the stage for us to move to more advanced topics on NLP and examine how the state of the art in NLP is now aiming to match human abilities as far as NLU is concerned.

© Navin Sabharwal, Amit Agrawal 2021
N. Sabharwal and A. Agrawal, *Hands-on Question Answering Systems with BERT*,
https://doi.org/10.1007/978-1-4842-6664-9_2

What Is a Neural Network?

A neural network is defined as a network of neurons that are connected to process information and perform actions specific to the task. To put it simply, human neurons have the capability to transmit and process information when they receive electrical signals at their synaptic endpoints. An artificial neural network (ANN) replicates this flow of information by transmitting the information across the network after getting triggered by an activation function. ANNs are divided into three types of layers: input layer, hidden layers, and output layer. Neural networks generally have one input and one output layer and multiple hidden layers, as shown in Figure 2-1.

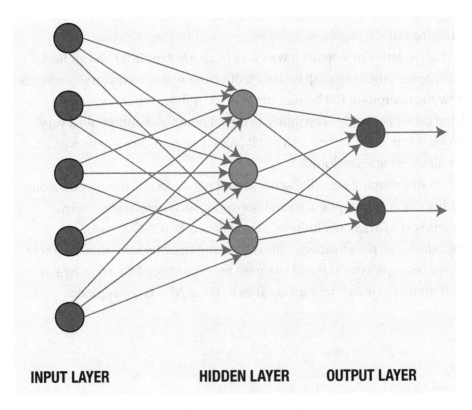

INPUT LAYER   **HIDDEN LAYER**   **OUTPUT LAYER**

Figure 2-1. Neural network

Building Blocks of Neural Networks

In this section we discuss the basic building blocks of neural networks and how these blocks can be combined to form a neural network.

Neuron

The neuron, which mimics the behavior of the human neuron, is the smallest unit of a neural network. It takes input, processes it, and sends output to other neurons that work as activation for others. A neuron can only be activated based on observations received from a previous layer.

Input Layer

As per Figure 2-1, the input layer takes processed data as input. Here, input can be pixels in the case of an image, or numbers from a vector representation of a sentence for text data or feature values. This layer is responsible for combining all features' values with some weight values (weight value defines how much importance is given to each feature). Once processed, the output from the input layer is fed into the next layer, a hidden layer, and at last into the output layer.

Hidden Layers

These layers are responsible for generation of features that are specific to a task. We can have any number of hidden layers between the input and output layer. Each layer consists of neurons that are responsible for performing actions specific to the task. This layer might either implement an activation function (i.e., Sigmoid, tanh), or can just do a weighted summation of all inputs from the previous layer. This layer therefore receives the input from previous layer and then performs a sum of the products of inputs with their corresponding weight value and applies an activation function to get output from this hidden layer. This information is then passed to the next hidden layer or output layer.

Output Layer

The output layer is the last layer in a neural network. It is responsible for gathering all information from the last hidden layer to output the final expected results. If you are working on classification model, then the last layer should have a number of nodes equal to number of classes or a single node in the case of a regression problem.

There has always been an open question of how we decide on the number of nodes in layers and nodes on each of the layers. There are no strict guidelines, but there are some recommendations you should consider while designing the a neural network architecture.

- Number of nodes in the input layer must be equal to the size of your input data point.

- Number of nodes in the output layer depends on the task the neural network is performing. As an example, for a classification task, the number of nodes should be equal to the number of classes and for regression it should be only one node.

- Number of hidden layers and nodes in each hidden layer is completely dependent on your target task. It is quite possible that one neural network will work perfectly for Task A but not for Task B.

- List all intermediate transformations you want to perform between the input and output layers.

- Number of nodes in a hidden layer should be greater than the number of nodes in the input and output layers.

- Number of nodes in the hidden layer should be a power of 2 (i.e., 2, 4, 8, 16, 32, etc.).

As an example, if you are building a sentiment model (or classification model) where the system will identify the sentiment of a user's feedback as positive, negative, or neutral, then the result from the output layer of the neural network is probability distribution across all classes (positive, negative, and neutral), as shown in Figure 2-2.

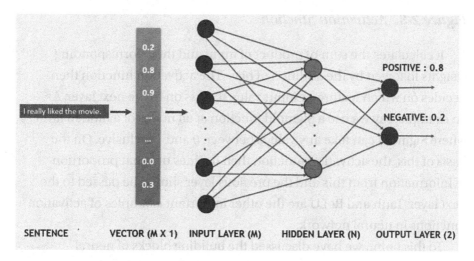

Figure 2-2. Neural network

Activation Function

Neural networks are used to solve complex nonlinear problems, which is not feasible with traditional linear models. The activation function in a neural network is the one that introduces nonlinearity in the system, as shown in Figure 2-3.

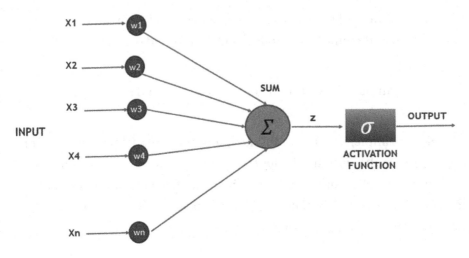

Figure 2-3. *Activation function*

It calculates the sum of product of input and their corresponding weights followed by the addition of bias. The activation function then decides on which feature or input value to pass on to the next layer. As an example, say we use a Sigmoid function at all nodes of a hidden layer where Sigmoid can take any values between 0 and 1 inclusive. On the basis of this, the activation function then decides on what proportion of information from this and the previous layer should be passed to the next layer. Tanh and ReLU are the other important examples of activation functions in neural network.

To this point, we have discussed the building blocks of neural networks. Next, we turn to the training of neural networks.

Neural Network Training

Neural network training is based on the concept of forward and backward propagation. In forward propagation, input data travels from neurons in the input layer to neurons in a hidden layer, followed by the application of relevant transformations using activation functions, and then finally to the neurons of the output layer to calculate the prediction value. In backward propagation, loss is calculated by comparing the actual and predicted value of input data. This error travels from neurons of the output layer to all neurons of the hidden layers. It is quite possible that neurons in the hidden layer receive only a fraction of an error component depending on their contribution to the neurons in the output layer.

When we talk about propagation of information either forward or backward it means the weights of edges connecting these neurons and values of these biases to neurons will be adjusted. Also, values of weights and biases are initialized randomly and the learning process finds optimal values of these model parameters accordingly.

Types of Neural Networks

Now that we understand how each of the neurons, weights, and activation functions together build a neural network, we can examine how each of these can be used in a different ways to achieve different results. Next we discuss a few types of neural networks.

Feed-Forward Neural Networks

Feed-forward neural networks (FNNs) can be best described as a unidirectional neural networks that do not have any feedback or loopback in their structure. The architecture of an FNN includes a number of hidden layers and a number of hidden units in each layer, as shown in Figure 2-4.

One of the reasons that this neural network is termed a feed-forward network is that there is no feedback between the layers during normal operations when the FNN acts as a classifier.

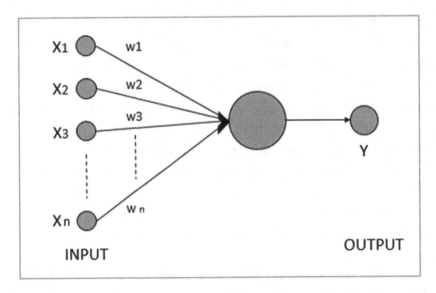

Figure 2-4. *Basic FNN perceptron diagram*

In an FNN, the perceptrons are arranged in layers. The first layer is responsible for taking in the input and the last layer produces the output. Because the middle layers do not have any connection with the external world, they are referred to as the hidden layers. The information is fed forward from one layer to the next as the perceptron in one layer is connected to every perceptron in the next layer. The perceptrons in the same layer, on the other hand, are not connected.

The FNN model uses a cost function during training. This cost function works on the difference between the approximation made by the model and the actual target value. Similar to ML algorithms, FNNs also use gradient-based learning for training purposes. The possible choices of cost function include quadratic cost, cross-entropy cost, exponential cost, and so on.

The output layer contains output units whose task is to provide the desired output or prediction. Both the choice of cost function and output units are tightly coupled together. There are various options for output units like linear units, sigmoid units, softmax units, and more.

FNNs are susceptible to noise in data and are easy to maintain, so they can have a huge scope of applications in fields like computer vision. FNNs can help in bringing out nonlinear relations between the input and output, so most of the multiclassification can be easily represented with the help of these networks.

Convolutional Neural Networks

A CNN is a deep learning algorithm. It takes in an input image and assigns weights and biases to various aspects in the image. There is comparatively less preprocessing in CNNs in contrast to other classification algorithms, as they possess the ability to learn filters and characteristics.

Similar to other neural networks, CNNs are also composed of neurons and have learnable weights and biases. The weighted sum is taken over several inputs received, and is then passed through the activation function along with an output. The CNN varies from other networks in the way that it operates over volumes. Here the input is not a vector; instead, the input is a multichannel image.

The CNN is used in image processing, as it captures the spatial and temporal dependencies in an image successfully by the application of relevant filters. The network understands the image in a better way as it performs better fitting to the image dataset while reducing the number of parameters involved.

Convolution is the combination of two inputs to generate an output. For CNNs, often this input is an image that is masked with a filter to generate the desired output features. When we talk in terms of spatially distributed data or matrixes as input, the filter that is chosen is often a set of weights to tune the input for the desired changes to generate a result. If a generic meaning is referred, then convolution stands as a dot product of two values to generate a third value. Figures 2-5 and 2-6 show how convolution works in CNNs.

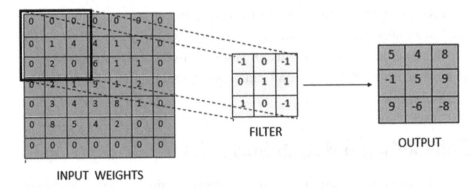

INPUT WEIGHTS

Figure 2-5. *Convolution example*

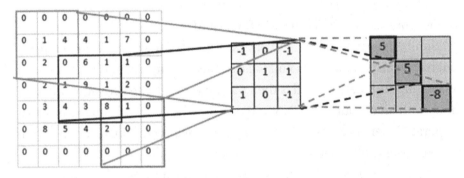

Figure 2-6. *Convolution example*

In the example shown in Figure 2-5, the input is in the form of a 5 × 5 matrix containing information from an image. We are going to apply a filter to yield the output. The input is padded with zeros to convert it into the below matrix. This padding is done to generate a spatial representation of the output in the desired dimensions.

In the given example, the stride or the step size is 2, meaning the filter moves two steps right horizontally and two steps down vertically. The first time the filter convolves the top left corner submatrix shown in blue:

$$(0x\text{-}1 + 0x0 + 0x\text{-}1 + 0x0 + 1x1 + 4x1 + 0x1 + 2x0 + 0x\text{-}1) = 5$$

24

The following steps fill in the rest of the output matrix. It is also noticeable here that the submatrices shown in blue, black, and red hold very different values, but after masking from the filter the blue and black submatrices yield an equal output and the red submatrix yields a far smaller value. This signifies how much an appropriate filter can help change the output feature matrix. For example, if an image has pixel values with huge contrast values, a proper filter can help tone down the image contrast.

CNNs play a crucial role by reducing images into a form that is easier to process while retaining the features that are important to obtain a good prediction. This ability of the CNNs makes them scalable to large datasets.

In a CNN, we have a convolutional layer that extracts the high-level features like edges from the input image. This layer is the building block of the CNN. It consists of a set of independent filters that are convolved with the image, giving us the feature maps. These filters are randomly initialized, and they become parameters on subsequent learning by the network.

Each of the neurons is connected to an input image's small chunk for a particular feature map. There is also parameter sharing in a particular feature map. All the neurons have the same connection weights in a particular feature map. Parameter sharing and local connectivity help in reduction of the number of parameters in the whole system and ensure better computational efficiency.

The concept of pooling makes CNNs differ from other neural networks. Pooling functions to reduce the spatial size of the representation progressively to reduce the number of parameters and amount of computation. The pooling layer operates independently on each of the feature maps.

After the pooling layer, the flattened output is fed to an FNN and then backpropagation is applied to every iteration of training, as shown in Figure 2-7. Over a series of iterations, the model is able to classify the images using the softmax classification technique.

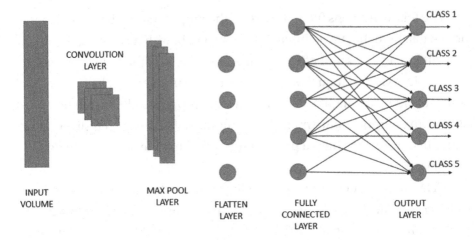

Figure 2-7. *CNN siagram*

Recurrent Neural Networks

Recurrent neural networks (RNNs) are neural networks that are designed to process continuous data or data that are presented as streams to benefit from the continuity of data. Often the data that are received at each hidden layer have an input from the previous layer's output as input to the current layer, along with a hidden input.

RNNs can prove to be very advantageous in cases with long input sequences where the requirement circles around maintaining the context of the same input, not affecting the size of the model being used. This makes NLP a natural fit application for RNNs, although historical information tends to fade over a long period of time, and also can slow down the process.

As we can see in Figure 2-8, the input is represented by X_t, which is the input to the network at time step t. For instance, X_1 can be one vector corresponding to a word in a sentence. The hidden state is represented by H_t at time t. It acts as the memory of the network. The value of H_t is computed on the basis of current input and the previous time step's hidden state:

$$H_t = f(U X_t + W H_{t-1})$$

The function f is a nonlinear transformation function like tanh or ReLU.

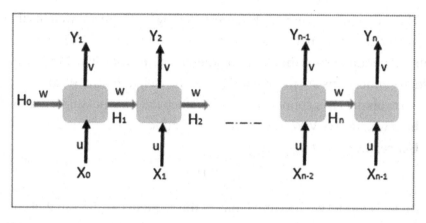

Figure 2-8. *RNN diagram*

In contrast to FNNs, RNNs make use of their internal state or memory to process the input sequences. All the inputs in an RNN are related to each other, unlike other networks where inputs are independent of each other. An RNN takes X_0 from the sequence of inputs and then it outputs H_0. This output, together with X_1, is the input for the next step. Hence, H_0 and X_1 form the input to the next step. Similarly, H_{t-1} and X_t form the input at time t. This way, the RNN remembers the context while training.

The current state is given by

$$H_t = f(H_{t-1}, X_t).$$

On applying the activation function,

$$H_{(t)} = \tanh(W H_{(t-1)} + U X_{(t)}).$$

- H is the single hidden vector.

- W is the weight at previous state.

- tanh is the activation function.

- U is the weight at the current input state.

The output of the network is represented by Y_t. There are weights that parameterize the connections from input to hidden layers. The weight matrix U parameterizes the input to hidden connections. The hidden to hidden connections are parameterized by the weight matrix W, and hidden to output layer connections are parameterized by the weight matrix V. All these weights (U, V, W) are shared across time.

Hence, the output is given by

$$Y_t = V\,H_t.$$

An RNN model enables modeling of the sequence of data so that each result of the sample can be assumed to be dependent on previous ones. There is also another advantage of RNNs, as they can even be used with convolutional layers to extend the effective pixel neighborhood.

The RNN model has a disadvantage when using tanh or ReLU as an activation function, as it fails to process long sequences. Training of an RNN is also a difficult task and there are problems of gradients vanishing and exploding.

Long Short-Term Memory

LSTM is one of the most widely used forms of RNNs. They are capable of learning long-term dependencies, and their default behavior to learn or remember information for long periods of time.

All RNN models have the form of repeating modules of neural networks that are chained. In standard RNNs, this repeating module will have a very simple structure, such as a single tanh layer.

LSTMs, on the other hand, also have this chain-like structure, but the repeating module has a different structure. Here, instead of a single neural network layer, there are four layers that interact in a very special way. The control flow of LSTM is similar to the RNN, as it processes the data and passes the information as the data propagates forward. The difference in the way LSTM works is that the cell allows the LSTM to keep or forget the information. In LSTM there is emphasis on cell state and the various gates. The cell state acts as a transport highway and transfers the relative information all the way through the sequence chain. The gates add or remove information as the cell state goes on the journey. The gates are different neural networks that decide which information is allowed in the cell state. The gates and cell state make LSTM distinctive among RNN models and further makes LSTM useful in various applications.

Encoders and Decoders

The encoder–decoder is an organization of RNNs for sequence prediction problems that often have a variable number of inputs, outputs, or both. The main purpose of the encoder–decoder initially was machine translation problems, but it has proven to be successful at related sequence-to-sequence prediction problems such as question answering and text summarization.

The encoder–decoder approach involves two RNNs, one to encode the input sequence and the other to decode the encoded input sequence into the target sequence. The encoding task is performed by the encoder and the decoding task is performed by the decoder. This encoder–decoder architecture is useful for various applications of sequence-to-sequence models like these:

- Chatbots

- Machine translation

- Text summary

- Image captioning

The Encoder–Decoder Architecture

The encoder–decoder architecture consists of two main components, the encoder and the decoder. Both these components are trained jointly at the same time. The architecture of the encoder–decoder is shown in Figure 2-9.

Figure 2-9. *Architecture of encoder–decoder*

The encoder takes the input and reads the entire input sequence, which it encodes into an internal representation. The encoder processes the input sequence and collects information from the sequence and then propagates it further. This fixed-length internal representation vector is known as the context vector. The intermediate vector ID is the final internal state produced from the encoder part of the model. This helps the decoder to make accurate predictions. The decoder is responsible for reading the encoded sequence from the encoder and thereby generating the output sequence.

Encoder Part of the Model

The encoder, which is responsible for converting the input sequence and encapsulating the information as the internal state vectors, is basically an LSTM or GRU (Gated Recurrent Unit) cell. Only the internal states are used; the outputs of the encoder are rejected, as shown in Figure 2-10.

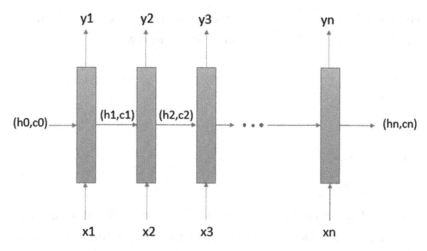

Figure 2-10. *LSTM for encoder*

To understand the working of the encoder part of the model, we focus on LSTM. In LSTM only one element is taken as input at a time. This implies that if we have a sequence of length m, then the LSTM takes m time steps to read the entire sequence.

- X_t is the input at time step t.

- h_t and c_t are internal states at time step t of the LSTM; for GRU there is only one internal state h_t.

- Y_t is the output at time step t.

Let's take an example of translation of a sentence in English into French.

English: It is a good day.

↓

French: C'est une bonne journée.

31

The English sequence shown can be considered a sentence containing five words. The inputs of the encoder X_t are as follows.

- $X_1 = $ It
- $X_2 = $ is
- $X_3 = $ a
- $X_4 = $ good
- $X_5 = $ day.

The LSTM will read the sequence word by word in five time steps. Each word X_t is represented as a vector using the word embedding. The word embedding converts each of the words into a vector of fixed length. The internal states (h_t, c_t) learn what LSTM has read until time step t. Here the LSTM will read the entire sentence in time step t = 5. The final state h_5, c_5 has the information of the entire input sequence, "It is a good day."

The output of the encoder is Y_t, which at each time step is the prediction of the LSTM. Because in machine translation problems we take the output of the entire input sequence, Y_t at each time step is discarded because it is of no use.

Decoder Part of the Model

The decoder works in a different way than the encoder. Its training phase and testing phase work differently, whereas the encoder model works the same way during the training and testing phases.

If we take the sentence language translation example presented earlier, just like the encoder, the decoder also generates the output sentence word by word. To generate the output "C'est une bonne journée," we need to add START_ at the beginning and _END at the end as delimiters of the output sequence so that the decoder recognizes the start and the end of the sequence. The decoder is basically trained to generate the output based on the information gathered by the encoder, so the initial states (h_0, c_0) of the decoder are set to the final states of the encoder.

The START_ is input so that the decoder can start generating the next word. The decoder is made to learn the end of the French sentence using _END. The loss is calculated on the outputs that are predicted from each time step and the errors are backpropagated through time to update the parameters of the model. In the testing phase, output produced at each time step is fed as input into the next time step and the end of the sequence is identified using _END.

Bidirectional Encoders and Decoders

In a bidirectional encoder–decoder architecture, the encoders and the decoders are bidirectional LSTMs. The last hidden state of the backward encoder initializes the forward decoder, whereas the backward decoder is initialized with the last hidden state of the forward encoder.

The bidirectional encoder is used when considering context information from the past and future. The sequence of input word vectors is fed to LSTM from forward and backward directions. The bidirectional decoder is also a bidirectional RNN that is made up of two separate LSTMs. One of the LSTMs decodes the information from left to right, whereas the other LSTM decodes in a backward direction from right to left. This bidirectionality in the RNN provides better performance.

For instance, we have to predict next word after "cloudy" in the sentence "The weather is cloudy; it might rain." The unidirectional LSTM will see "The weather is ..." and will try to predict the next word using this context only. When using bidirectional LSTM we will be able to see more information.

Forward LSTM: "The weather is ..."

Backward LSTM: "... it might rain today."

Hence using the information from the past as well as the future makes it easier to predict the word "cloudy," as the network will better understand the next word.

Transformer Models

The transformer is a novel architecture with the aim of solving sequence-to-sequence tasks while handling long-range dependencies. The transformer maintains sequential information in a sample just as RNNs do. If we take a high-level look at the transformer model, it basically is like a single black box in machine translation application that takes a sentence in one language as an input and outputs the translation of the sentence, as shown in Figure 2-11.

Figure 2-11. *The transformer as a black box*

Model Architecture

The transformer has an encoder–decoder structure using stacked self-attention and fully connected layers for both the encoder and decoder. The transformer consists of components like encoders, decoders, positional encoding, and attention. There is a stack of encoders and decoders. Each of the encoders is very similar to the other encoders, as they have the same architecture. Decoders, too, share this property and are similar to each other in the transformer, as shown in Figure 2-12.

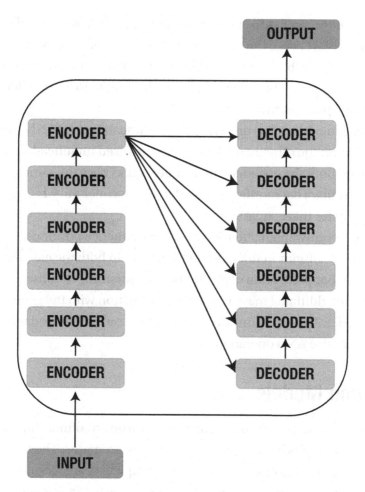

Figure 2-12. *Encoder–decoder stacks in transformer*

The encoder comprises a stack of identical layers. Every layer further consists of two sublayers. The first layer is for a multihead self-attention mechanism. The second layer, on the other hand, is a simple, fully connected feed-forward network. In between each of the sublayers, residual connections are employed along with layer normalization. The input flows through the self-attention layer in the encoder and helps the encoder to look at other words in the input sequence while encoding a specific word.

35

The decoder, similar to encoder, comprises a stack of identical layers. It also has sublayers like the encoder, but has one more sublayer. This third sublayer is responsible for performing multihead attention over the output of the encoder stack. These layers help the decoder to focus on relevant parts of the input sentence only.

The encoder block has one layer of a multihead layer of FNN, and the decoder, on the other hand, has an extra masked multihead attention mechanism. Both the encoder and decoder stacks have the same number of units. That number of encoder and decoder units is basically a hyperparameter that can be varied.

To prevent unwanted attention to out-of-sequence positions, masking is used before softmax in the self-attention layer in both the encoder and the decoder. For prevention of positions from attending to subsequent positions, an additional mask is used in conjunction with the general mask in the decoder. These two masks in the decoder can be blended with the help of a bit-wise AND operation.

Attention Models

Attention is the output vector of a dense layer using a softmax function. It enhances the results by plugging it into suitable scenarios. The translation mechanism is used to rely on reading of a complete sentence and compressing all information into a fixed-length vector. In this situation if we have a sentence with hundreds of words that are represented by several words, there will surely be a loss of information or inadequate translation.

Attention is able to partially fix this problem. It is responsible for allowing the machine translator to look over all of the information that the sentence contains. Thereby, the proper word is generated according to the current word and the context. Attention also provides the ability to focus on local or global features by allowing the translator to zoom in or out.

Why Is Attention Required?

As sentences consist of different numbers of words, an RNN is naturally introduced to model the conditional probability among words. In a probabilistic language model, the focus is to assign a probability to a sentence using a Markov assumption.

$$P(w_1w_2 \dots w_n) \approx \Pi P(w_i \mid w_{i-k} \dots w_{i-1})$$

The translation works on input and output of variable lengths with this encoder–decoder model. This is adopted while the basic RNN cell is changed to an GRU or LSTM cell and ReLU replaces the hyperbolic tangent activation.

The discrete words are mapped to dense vectors for computational efficiency with the help of an embedding layer. These embedded words are then fed sequentially to the encoder. As the information flows from left to right, every word vector is learned according to all the previous inputs, not just the current word. Once the sentence is read completely, output is generated by the encoder. There is also a hidden state that is generated by the encoder for further processing. The decoder uses this hidden state from the encoder and generates the translation words sequentially.

How Attention Works

The attention mechanism is basically a context vector that is plugged into the encoder–decoder architecture within the gap between the encoder and decoder. This context vector takes all the encoder cell's output as input and then computes the probability distribution of source language words for each of the words that the decoder wants to generate. The decoder is able to capture global information, not just infer on the basis of one hidden state. This attention mechanism helps the decoder to capture a wider perspective.

If we look into how a context vector is built, it is basically quite simple. For every fixed target word, we generate scores for each of the encoder states by looping over all the states of the encoder and comparing the target to the source states. Then softmax is used to normalize all the scores. We now obtain the probability distribution conditioned on the target states. Finally, to make the context vector easy to train, the weights are introduced. Once we get the context vector, the attention vector can be easily computed using the context vector, the attention function, and the target word.

Types of Attention Models

There are three types of attention models: global and local attention, hard and soft attention, and self-attention. Let's examine each in turn.

Global Attention Model

In the global attention model, inputs from every encoder state and decoder state prior to the current state is taken into consideration for computation of the output. The context vector here is obtained by taking product of global aligned weights and each of the encoder steps. This is then fed to the RNN cell to obtain the decoder output.

Local Attention Model

The local attention model varies from the global attention model as few positions from the encoder are used for calculation of the aligned weights. Local attention models are further of two types: monotonic alignment and predictive alignment.

Hard and Soft Attention Model

The soft attention model is similar to the global attention model. The hard attention model differs from the local attention model in that the local model is almost differential at every point, whereas the hard attention model is not. The local attention model can be considered a blend of hard and soft attention.

Self-Attention Model

The self-attention model relates different positions of the same input sequence. Self-attention can theoretically adopt any score functions conditioned that the target sequence is to be replaced with the same input sequence.

Conclusion

In this chapter, we have discussed about various Neural networks in NLP domain. Now that we have covered different types of neural networks, we turn our attention to how we can use BERT.

CHAPTER 3

Introduction to Word Embeddings

NLP tasks such as document classification, sentiment analysis, clustering, and document summarization require processing and understanding of textual data. Implementation of these tasks depends on how data are being processed and understood by AI systems. One way of doing this is to convert textual representation to a numerical form using some statistical methods such as term frequency-inverse document frequency (TF-IDF), count vector, and so on, but these methods do not consider the meaning of a sentence and only deal with the occurrence of words in sentences.

Over the course of time, several semantic methods such as parse trees, contextual grammar, ontologies, and others have been developed, but these methods would require a great amount of human effort to prepare labeled training data. In the last few years, widespread availability of computing capacity has made it possible to use neural network–based methods for these tasks.

One-Hot Representation

One-hot representation is one of the most common and basic methods for representation of text. It involves representation of a word using binary encodings (i.e., 0 and 1). It can also be used for representation of categorical attributes.

© Navin Sabharwal, Amit Agrawal 2021
N. Sabharwal and A. Agrawal, *Hands-on Question Answering Systems with BERT*,
https://doi.org/10.1007/978-1-4842-6664-9_3

As an example, assume that a dataset has color as one of the features, with three possible values: red, blue, and green. Therefore, this feature will be converted to three new columns, one for each color value, as shown here.

	RED	BLUE	GREEN
1	1	0	0
2	0	1	0
3	0	0	1

As an example, the first data point has value of 1 in the RED column and 0 in the others. This means initially that this data point has value for the color column of RED. Red is represented as [1 0 0] in one-hot encoding, for blue 1 occupies the second position, and for green 1 is in the third position. This is a three-dimensional vector. The one-hot representation expands the feature vector as each category of the color is itself a feature now.

Now, when we talk about one-hot encoding for textual sentences, it does not care about order of occurrence of words in sentences and actually ignores semantic meaning of words. This approach works best in scenarios where the corpus is smaller and some traditional NLP methods need to be used.

As an example, to represent a text sentence using one-hot representation, following these steps.

1. Count the total number of unique words present in corpus.

2. Assign 0 or 1 depending on if the word is present in the sentence or not.

Consider the sentence "The sky is clear today." The vocabulary includes words such as The, sky, is, clear, and today. It forms a five-dimensional vector if represented in one-hot representation as shown in Figure 3-1.

```
The: [1 0 0 0 0]

sky: [0 1 0 0 0]

is: [0 0 1 0 0]

clear: [0 0 0 1 0]

today: [0 0 0 0 1]
```

Figure 3-1. *One-hot representation*

Count Vector

In the previous section, we saw how one-hot representation of a sentence is generated on the basis of occurrence of words, not on the basis of their frequency of occurrence. The count vector for an individual sentence is generated on the basis of the number of times a particular word occurs in the sentence. The unique words in the corpus form the vocabulary.

As an example, consider these two sentences.

> Sentence 1: The blue bird is flying in the clear blue sky.

> Sentence 2: The sky is clear today.

This corpus has two sentences and the vocabulary set [bird, blue, flying, is, in, the, sky, clear, today] contains nine terms. For every word in the vocabulary set, its frequency of occurrence in a sentence is determined. A count vector corresponding to that sentence is thus formed. From these count vectors that represent the sentences we get our count matrix. The count matrix for the two example sentences is shown here.

	is	bird	blue	flying	In	the	sky	clear	today
Sentence 1	1	1	2	1	1	2	1	0	0
Sentence 2	1	0	0	0	0	1	1	1	1

The rows in the matrix are representative of the sentences and the columns signify the word vector for the corresponding word in the matrix. The size of the matrix is S × T, where S is the number of sentences and T is the number of terms or words.

Count vector representation of sentences helps us to achieve several tasks, including these:

- Determining similarity between sentences

- Identification of relevant documents for a query

- Document summarization

As an example, we are showing how similarity between sentences can be computed mathematically. The count vector for Sentence 1 is [1 1 2 1 1 2 1 0 0] and the count vector for Sentence 2 is [1 0 0 0 0 1 1 1 1]. The cosine similarity can be computed using the following mathematical expression.

$$\text{Sim} = \frac{x \cdot y}{\|x\|\|y\|}$$

Here, x and y are the two count vectors. $\|x\|$ is the Euclidean norm of vector x.

So,

$$x.y = 1x1 + 1x0 + 2x0 + 1x0 + 1x0 + 2x1 + 1x1 + 0x1 + 0x1 = 4$$

$$\|x\| = \sqrt{1^2 + 1^2 + 2^2 + 1^2 + 1^2 + 2^2 + 1^2 + 0^2 + 0^2} = 3.60$$

$$\|y\| = \sqrt{1^2 + 0^2 + 0^2 + 0^2 + 0^2 + 1^2 + 1^2 + 1^2 + 1^2} = 2.24$$

$$\text{Sim} = 4/ (3.6 * 2.24) = 0.49$$

This calculation means these two sentences are similar to each other with a similarity score of 49% (0.49). The similarity value will always lie between 0 and 1, where 1 indicates maximum similarity and 0 means no similarity.

TF-IDF Vectorization

One-hot representation and the count vector method are the most basic methods that do not actually consider the importance of a particular word in a sentence and in a corpus. For some NLP projects such as search engines, it is very important to know about the importance of words in a query to words in documents in your corpus to determine the relevancy of documents to that query. Some English words that occur frequently (e.g., "is," "the," "a," etc.) will be present in all the documents. Even though their count is higher, they are not useful when performing NLP-related tasks. To overcome this drawback of count vectors, TF-IDF is used. TF-IDF is one of the most popular techniques used in various applications as it is able to weight the words that appear more frequently in general.

In TF-IDF we form vocabulary in a way similar to the previous method. The vocabulary consists of unique words across the corpus. Now term frequency is computed for every word in the vocabulary set. Term frequency (TF) of a word or term t corresponds to the count of all its occurrences in a document d to the number of terms in the document.

TF = (Number of times term t appears in a document / Number of terms in the document)

We compute the inverse document frequency (IDF) by calculating the count of documents in which that term is present. IDF tells us about how much information a term or word gives. It tells you if a word is common across all documents or not. IDF is the log value of the ratio of total number of documents to the number of documents in which a term t has appeared.

IDF = log(N/n) where N is the total number of documents in the corpus and n is the number of documents in which a term t has appeared

TF-IDF is the product of TF and IDF.

TF-IDF (t, document) = TF (t, document) * IDF(t)

For example, let's use the previous two example sentences, where Sentence 1 and Sentence 2 correspond to two documents.

TF of word "blue" in Sentence 1 = 2

IDF = log (2/1) = 0.3

TF-IDF of word "blue" in Sentence 1 = 2*0.3 = 0.6

Similarly, the TF-IDF value of the word "is" in Sentence 2 is 0, as its IDF score is 0. This signifies that the word "is" does not have any importance because it is common and present across all the documents.

Methods like TF-IDF, count vector, and one-hot encoding are easy to compute, but they do not capture semantics (or order of occurrence of words) in the document. The words or the sentences represented using these methods do not provide any contextual information. Even though many NLP tasks can be performed using them, the overall results are mediocre, especially when the training data are sparse. A better technique is therefore required that can capture useful language information and at the same time boost generalization and performance for pretty much any NLP problem.

In next section, we are going to discuss one such approach, word embedding, where vector representation of words contains contextual information.

What Is Word Embedding?

Word embedding is a type of word representation where the words are embedded into vectors of real numbers. The embeddings can be generated through approaches like neural networks, probabilistic models, or dimension reduction on a word co-occurrence matrix, as shown in Figure 3-2. They enable the words with similar meanings to be understood by the ML algorithms.

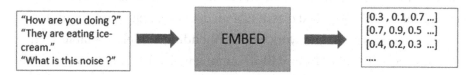

Figure 3-2. *Word embedding*

Word embeddings are generally low-dimensional (usually 50–600 dimensional) and dense representations of words or sentences as compared to one-hot representation. When using one-hot representation, the feature vectors increase with the size of the vocabulary set. Word embeddings, on the other hand, are more efficient. They have the ability to generalize. Semantically similar words are more likely to have similar vector representations. Hence, these vectors will give you more relevant results when used with NLP tasks such as document summarization, sentence or document similarity, and so on, as compared with one-hot representation.

Word embedding is also known as a distributed represented or distributed semantic model or semantic vector space. The word "semantic" here highlights the significance of word embedding as it aims to categorize words with similar meanings together. For example, sports like tennis, football, and swimming should be placed close, whereas words related to animals would be far away from these words. In a broader sense, word embedding will create vector representations of words related to sports that will be placed far away from vector representations of words for animals. The main objective is to have words with similar contexts occupy the closest spatial positions.

Embeddings are generally the vectors that are indeed the representation of words in lower dimensions. Neural networks are also currently being used to generate embeddings of words. It improves the ability to learn or generalize representations from the last set of textual data. Neural network models can learn resourceful traits about words in a vocabulary set while reducing the dimensionality of the text data. Word embeddings prove to be very useful in NLP tasks, text classification, document clustering, and so on. There are various neural network word embedding models available such as Word2vec, GloVe, ELMo, and BERT, among which BERT has proven to be best to this point for state-of-the-art NLP tasks.

Different Methods of Word Embedding

There are different methods to generate embedding of words and they differ by their implementation approach. Next we discuss some of them in detail.

Word2vec

Word2vec is a shallow, two-layered neural network technique of word embedding in which the words are represented in vector space. A neural network with only a hidden layer between the input and output layer is termed a shallow neural network. Word2vec is a two-layer network with an input layer, one hidden layer, and an output layer. It takes a text corpus as input and gives a set of vectors as output. The feature vectors represent the words of the corpus. The vectors that represent the words are known as neural word embeddings.

This vector representation maintains a semantic relationship between words in the document or corpus. Words with similar meanings will be located quite close to each other in vector space and the dissimilar words are located far away. The semantic relationship is achieved as Word2vec reconstructs the linguistic context of words. The linguistic context can be understood as the main objective of the sentence. For instance, in the sentence "What date is today?," a person wants to know today's date, which is actually the context of the sentence. The main context can be disclosed by the words and the sentences surrounding the language. Association of a word with other words can be guessed accurately with the help of Word2vec when given enough text corpus.

Word2vec is able to train the words against their neighboring words in the input text data. It can be implemented in two ways: continuous bag of words (CBOW) and skip gram. These are the two implementations of Word2vec that are used to create word embedding representations. In CBOW, context is used to predict the target word, whereas in skip gram, a word is used to predict the target context.

Continuous Bag of Words

The CBOW architecture tries to predict the target word using the context window words. The center word or target word is predicted with help of the surrounding words or the source context words. The Word2vec models are unsupervised models, which implies we need to provide only the input corpus without any additional information about the output. To get the CBOW word embeddings, the model follows a supervised classification methodology such that it takes the corpus as the input X and predicts the target word Y.

The input to the neural network will be the sum of one-hot encoded vectors of the context words in the given window size. The logarithmic loss function will be used as the loss function.

$$-\frac{1}{N}\sum_{i=1}^{N} y_i \cdot log\big(p(y_i)\big) + (1-y_i) \cdot log\big(1-p(y_i)\big)$$

A softmax function is being used as an activation function in the last layer. This will provide you a probability distribution across all words. The equation for softmax function is shown here:

$$\sigma\left(x_j\right) = \frac{e^{x_j}}{\sum_i e^{x_i}}$$

For example, consider the sentence "This beautiful painting belongs to Queen Elizabeth." Here are some of examples of training data considering the context window size of 2.

- [(painting, belongs), Queen Elizabeth]

- [(This, beautiful), painting].

The input layer will have a one-hot representation of context words (i.e., "This" and "beautiful") and the output layer will show probability distribution across all words in the corpus where the probability score for the word "painting" will be highest one.

Figure 3-3 shows the architecture for both variations of Word2vec embedding.

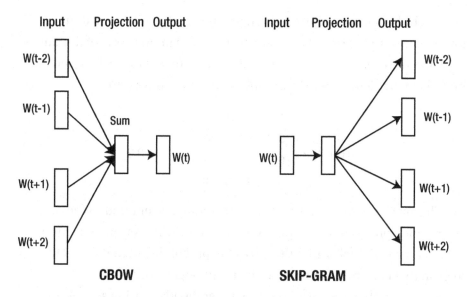

Figure 3-3. *Architecture of CBOW and skip gram models*

Skip Gram Model

The skip gram model is an unsupervised learning technique that finds the most relevant words around the target word. In this context, words are predicted using the target word. It is the reverse of the CBOW approach. Here the target word is the input and the context words are the output. It is a comparatively difficult technique, as more than one context word is to be predicted. As seen in the skip gram model architecture (Figure 3-3), the input is the target word $W(t)$ and the output is the vector representation of context words.

This is generally computed as per the following method. The dot product between this input vector and the weight matrix is obtained by the one hidden layer. Similarly, in the output layer the dot product is computed between the output vector of the hidden layer and the output layer's weight matrix. Then to find the probability of the words to in context of $W(t)$ is calculated using the softmax activation function.

The hidden layer is the weight matrix where the rows contribute toward the output words. For instance, if the weight matrix is of dimension 4 × 4 and the input is a one-hot encoded word, then a row will be selected from the matrix corresponding to the one in the input vector.

$$[0\ 0\ 1\ 0] \times \begin{pmatrix} 10 & 11 & 4 & 9 \\ 3 & 2 & 6 & 16 \\ 6 & 15 & 3 & 2 \\ 5 & 14 & 3 & 8 \end{pmatrix} = [6\ 15\ 3\ 2]$$

This word vector is obtained after the hidden layer is fed to the output layer, which produces an output between 0 and 1. The output layer is a softmax regression classifier that gives the probability of the output word to be in that context position near the input target word.

Figure 3-4 shows an example of word embedding using skip gram.

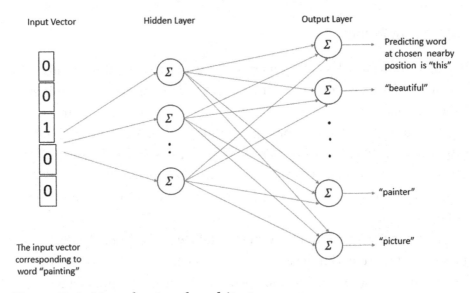

Figure 3-4. *Neural network architecture*

GloVe

GloVe (Global Vector) is an unsupervised technique that is used for the vector representation of the words in the global corpus. The word vectors are obtained by taking into consideration both global and local statistics of a corpus. The local statistics correspond to the local context information of words, whereas global statistics are captured by word cooccurrences.

Even though Word2vec performance was quite satisfactory, there remained a need for a better approach, as Word2vec only considers the surrounding words, which might sometimes fail to capture useful relationships of a word with other words. The semantics learned in case of Word2vec is only dependent on the local information and is affected by the neighboring words. In GloVe, on the other hand, the meaning of a word can be obtained with the help of the structure of the whole corpus. This constitutes word frequency and cooccurrence count. This model mainly relies on the intuition that word-to-word cooccurrence probabilities can contribute to encoding some form of meaning that gives it an extra benefit over Word2vec.

GloVe trains on these aggregated global word–word cooccurrence statistics and minimizes the least square error. This results in the meaningful linear substructure of a word vector space. For example, "man" and "woman" are similar in the context that they both describe human beings, but these two words are also opposites. To capture as much of the meaning specified by the two words as possible, we need a larger information corpus. The discrimination between the two words is based on gender, which can be specified by other word pairs like husband–wife, brother–sister, and so on.

Sentence Embeddings

Word embedding generates a vector representation of words by considering only neighboring words, not other sentences. To capture the relationship between sentences, sentence embedding is the best approach. These are vector representations of the sentences in a document. Sentence embedding models are essential, as they are capable of capturing contextual information that word embedding models fail to capture. As discussed previously, the word embeddings represent the meaning of words in a sentence or conversation. They are a representation of words in an N-dimensional vector space. These methods often tend to neglect necessary information, however, as explained in the example that follows.

Two sentences can have identical representations but entirely different meanings. For instance:

Sentence 1: The sky is clear not cloudy today.

Sentence 2: The sky is cloudy not clear today.

Here, Sentence 1 and Sentence 2 have similar representations, but their meanings are entirely different. Word embeddings won't be able to differentiate between these two sentences because the vector representation of words present in these sentences would almost be same. Sentence embeddings can be used to accomplish this differentiation.

When working with textual data in the ML pipeline, we do come across the need to compute sentence embeddings so that we are able to embed full sentences into a vector space. Sentence embeddings can capture semantic similarity or relatedness between sentences, then paragraphs, then documents.

A sentence embedding for a sentence might look like: this

"The bird is flying in sky." – [0.1, 0.7, 0.4, …]

To generate sentence embedding for a sentence, the most basic approach will be to perform an average of word embeddings of all words present in that sentence.

A weighted average of the word embeddings can be used to obtain the sentence embeddings and reduce the dimensionality. In addition to this method, other methods such as Universal Sentence Encoder and ELMo have been introduced that have turned out to be very useful for NLP-related tasks.

ELMo

ELMo (Embeddings from Language Models) is a deep contextualized word embedding. It was developed in 2018 by the Allen Institute of AI. ELMo uses a deep bidirectional LSTM model to create word representations. The internal states of the two-layer bidirectional language model compute the embeddings. It is able to capture the changing contextual meaning of words in the sentences, as shown in Figure 3-5.

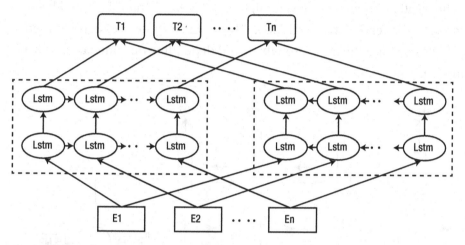

Figure 3-5. *ELMo architecture*

ELMo analyzes the words within the context in which they are used, unlike Word2vec. It does not create vectors for a dictionary of words; instead, vectors are created by passing text through the deep learning model. The representation of a word is dependent on the entire sentence corpus that is passed to the model. It does not use a fixed embedding for each word; instead the entire sentence is looked at before assigning embedding to a word. It is able to understand the meaning of a word along with the context in which it is found. Hence it is able to capture meaning along with the contextual information. This contextual information related to a word might vary depending on the sentences in which the word is used. This gives it an advantage over Word2vec and GloVe. The pretrained language embeddings, when added to existing models, improve the state of the art across NLP problems.

ELMo is character based: It takes characters as input instead of words, which enables it to compute meaningful representations for words not seen during training. When trained on a large dataset, it is also able to learn the language patterns, which is beneficial for tasks related to NLU, like determining the next word in a phrase. For example, in the phrase "The weather is cloudy today, it might … ," the word "rain" is more likely to appear instead of the word "dog." The model is quite useful in scenarios like these to find the most probable word depending on context, as shown in Figure 3-6.

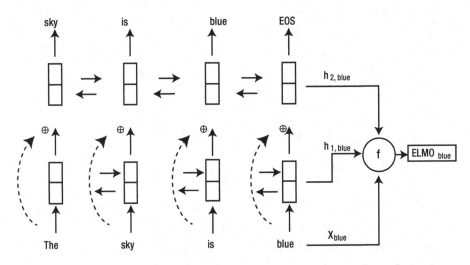

Figure 3-6. *ELMo specific representation of the word "blue"*

The ELMo model is a fairly sophisticated neural language model that seeks to compute the probability of a word, given some prior history of words seen. The ELMo architecture (refer to Figure 3-5) has a two-layer bidirectional LSTM as its backbone. This two-layer bidirectional LSTM model helps the model to understand the next word as well as the previous word in a sentence. The first and second layers are connected by residual connections that can skip one or multiple layers as the layer feeds into the next layer and directly into layers hops away. They are used to make deeper networks easier to optimize. In the ELMo language model, each token is converted into an appropriate representation using character embeddings. We get these character-level embeddings using a one-dimensional CNN to obtain a numerical representation of a word. This allows a valid representation even for words not in the vocabulary set. This is then passed through the convolutional layer using various numbers and types of filters. Finally, before passing as the input to the LSTM layer, it is passed through a two-layer highway network. This highway network enables smoother transfer of information through the input. These transformations to the input token allow selection of the morphological features, n-gram features, and more. This helps to build a powerful representation of sentences.

Let us assume we are looking at the ith word in our input. Taking Figure 3-5 as a reference, the ELMo representation of the word "blue" is the combination of the transformed word representation x_i as well as the output of two bidirectional representations h_{1i} and h_{2i}. The function f performs the following operation on input.

$$\text{ELMo}_i{}^{task} = \gamma i \cdot (s_0{}^{task} \cdot xi + s_1{}^{task} \cdot h_{1,i} + s_2{}^{task} \cdot h_{2,k})$$

Here, γi and s_k are the weighting factors that are learned during the task-specific model. So, when we use ELMo, we freeze the weights and then concatenate the ELMo i task for each token to the input representation.

Universal Sentence Encoder

Universal Sentence Encoder was introduced recently, and it has become one of the most popular pretrained models for sentence embeddings. It is able to convert sentences into vector representations. This versatile sentence embedding model can learn rich semantic information and thereby use transfer learning where sentence representation from other tasks can be learned by retraining the last layer of the architecture.

This sentence encoder model can be used for a wide variety of NLU tasks. The transformer network used by the encoder is trained on a large, varied data corpus. The input text, which can be a sentence, phrase, or short paragraph, is encoded into a high-dimensional vector. Here, input length can be variable but the output is a 512-dimensional vector. This enables generation of sentence embedding for a broad range of downstream tasks like text classification, clustering, semantic similarity, and more.

Several versions and implementations of these models that have been trained by Google using Tensorflow are available on the TensorFlow hub for ML engineers to consume, including these.

- universal-sentence-encoder-large

- universal-sentence-encoder-lite

- universal-sentence-encoder-multilingual

- universal-sentence-encoder-multilingual-large

- universal-sentence-encoder-multilingual-qa

Bidirectional Encoder Representations from Transformers

Bidirectional Encoder Representations from Transformers (BERT) has been introduced by researchers at Google. The bidirectional transformer for language modeling makes BERT popular in a variety of NLP tasks as well as question answering. This makes it different from the previous models where sequences are taken in one direction only, either left to right or right to left.

The bidirectional encoder takes two sequences for encoding, one of which is the normal sequence and the other one is the reverse of it. It consists of two encoders for encoding the two sequences. For the final output, both encoding results are considered. The bidirectional training of language models gives them deeper insight into the context of language. It is indeed important for understanding the meaning of text, as shown in Figure 3-7.

For example, consider the following two sentences:

> Sentence 1: I got scared on seeing a **bat** flying in my room.
>
> Sentence 2: The player held the **bat** firmly while smashing a ball with it.

Here, the word "bat" has a different meaning in the two sentences, depending on the language context. This is better understood if we approach the sentence from both directions. If we move in only one direction, we might miss useful information and might not correctly obtain the meaning. BERT considers both the preceding and following context, which reduces the chances for an error before making any prediction. Information is gathered from both directions while training and the context from both directions is jointly conditioned in all the layers.

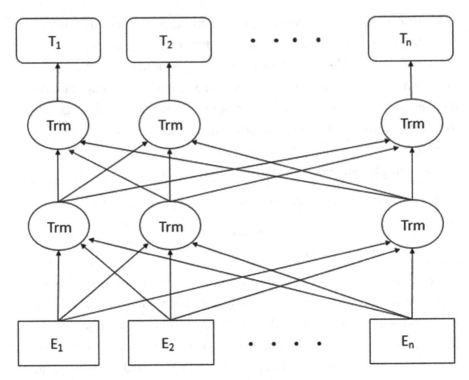

Figure 3-7. BERT architecture

The pretrained BERT model can be used for various state-of-the-art tasks by just modifying the output layer. It does not require any task-specific architectural change. It uses the transformer to grasp the relationship of a token or word in the text. A transformer includes an encoder and a decoder. The encoder reads the input text and the decoder helps in generating predictions for a task. The transformer encoder is able to read the entire sequence of words at once instead of reading sequentially from left to right. This makes the model bidirectional and allows it to learn the context of a word or token from both left and right side of it. The sequence of tokens that are input to the transformer are embedded into vectors and then vectors are further processed in the neural network. The output of the network is a sequence of vectors corresponding to the input tokens, as shown in Figure 3-8.

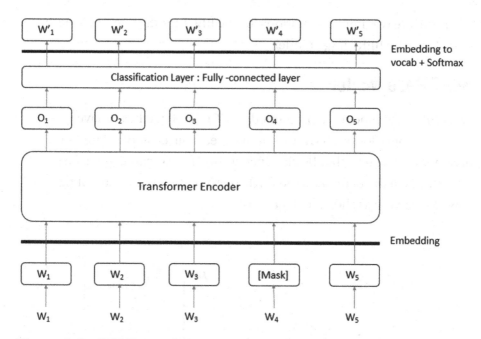

Figure 3-8. *BERT transformer*

BERT uses two strategies to surpass unidirectional constraints. BERT is pretrained on these two NLP tasks: masked language modeling (MLM) and next sentence prediction (NSP). MLM assists in pretraining the bidirectional transformer by randomly masking tokens from the input text while the NSP task jointly pretrains text pair representations. BERT minimizes the combined loss function for both the tasks during training.

To use BERT, two stages are to be followed:

1. **Pretraining:** In this step the model is trained on unlabeled data over different pretraining tasks.

2. **Fine-tuning:** The BERT model is initialized with the pretrained parameters followed by fine-tuning using the data from the downstream task, which could be classification, question answering, and so on.

There are two implementations for the BERT model, the BERT base model and the BERT large model.

BERT Base Model

The BERT base model is a pretrained BERT model that has 12 layers or transformer blocks, 768 hidden units in each layer, and 110 million parameters. It can further be classified as BERT base-cased and BERT base-uncased depending on the English text (cased or uncased) it has been trained on, as shown in Figure 3-9.

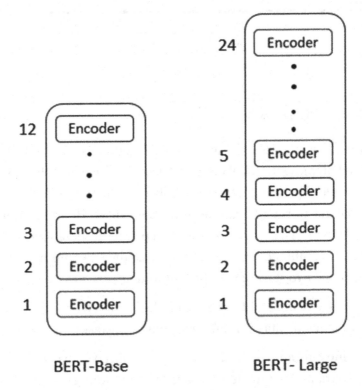

Figure 3-9. *BERT base and BERT large models*

BERT Large Model

The BERT large model is a pretrained BERT model that has 24 layers or transformer blocks, 1,024 hidden units in each layer, and 340 million parameters. It can also further be classified as BERT large-cased and BERT large-uncased. This model requires significantly more memory than BERT base.

Conclusion

This chapter covered word embedding, sentence embeddings, and their different methods of implementation, such as Word2vec, GloVe, Universal Sentence Encoder, and so on. We have also discussed BERT and its variations (i.e., base and large models). In the next chapter, we will look deeper into BERT and its different implementations.

CHAPTER 4

BERT Algorithms Explained

This chapter takes a deep dive into the BERT algorithm for sentence embedding along with various training strategies, including MLM and NSP. We will also see an implementation of a text classification system using BERT.

How Does BERT Work?

BERT makes use of a transformer to learn contextual relations between words in a text. A transformer has two mechanisms—an encoder and a decoder—but BERT only requires the encoder mechanism. BERT uses a bidirectional approach and reads the text input sequentially, which allows the model to learn the context of a word based on its surrounding words. The input to the encoder is a sequence of tokens that are embedded into vectors. The vectors are then passed into the neural network and an output sequence of vectors is then generated corresponding to the input. The output vector for a word is dependent on the context in which it occurs. For example, the vector for the word "like" in the sentence "He likes to play cricket" would be different than the vector for the same word in the sentence "His face turned red like a tomato."

© Navin Sabharwal, Amit Agrawal 2021
N. Sabharwal and A. Agrawal, *Hands-on Question Answering Systems with BERT*,
https://doi.org/10.1007/978-1-4842-6664-9_4

This procedure involves text processing steps before even starting the model-building phase. The next section discusses the text processing steps used in BERT.

Text Processing

There is a specific set of rules for representing the input text for the BERT model. This, too, is responsible for better functioning of the model. If we look into the embeddings, the input embedding in BERT is a combination of the following three types of embeddings.

- **Position embeddings:** Positional embeddings are used to learn the information of order in the embeddings. As in transformers the information related to order is missed, positional embeddings are used to recover it. For each of the positions in the input sequence, BERT learns a unique positional embedding. With the help of these positional embeddings, BERT is able to express the position of words in a sentence as it captures this sequence or order information.

- **Segment embeddings:** BERT also learns unique embedding for the first and second sentences to help the model distinguish between them. It can also take sentence pairs as inputs for tasks like question answering.

- **Token embeddings:** For every token in the WordPiece token vocabulary, token embeddings are learned. The WordPiece token vocabulary contains subwords of words in the corpus. As an example, for the word "Question," this vocabulary set will include all possible subwords of "Question," such as ["Questio", "Questi"...], and so on.

Figure 4-1 shows an example of sequences of embeddings in BERT.

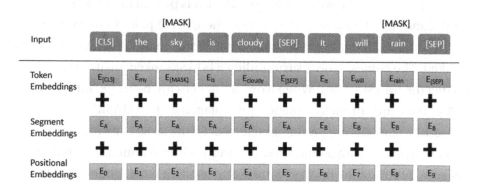

Figure 4-1. *BERT embeddings*

The input representation of a given token is constructed by summing the token, segment, and position embeddings. This makes it a comprehensive embedding scheme that contains a lot of useful information for the model.

For an NLP task where the job is to predict next word in a sentence, if we go with a directional approach, it has some limitations. However, BERT provides two strategies to learn contextual information: MLM and NSP. During training in BERT, both of these tasks will be trained together. When using these two strategies, the model tries to achieve the goal of minimizing the combined loss function.

Masked Language Modeling

BERT is a deep bidirectional model that is more powerful than a left-to-right model or the shallow concatenation of a left-to-right and a right-to-left model. The BERT network can effectively capture information from both the right and left context of a token. This goes from the first layer itself and all the way through to the last layer. Previously, language models were trained on left-to-right context, which made them susceptible to information less. Even though the ELMo model greatly improved on the existing techniques using the shallow concatenating of the two LSTM language models, that wasn't enough. BERT has proven to be more significant than the existing techniques where MLM plays a crucial role.

In a masked language task, some of the words in text are randomly masked. The context words surrounding a [MASK] token are used to predict the [MASK] word. When word sequences are being fed into BERT, 15% of the words in each sequence are replaced with a [MASK] token. These 15% of words are randomly selected. Of these, 80% are masked, 10% are replaced with a random word, and 10% are retained. This is done because if 100% of the masked words were used then the model wouldn't necessarily produce good token representations for nonmasked words. The model performance is improved, as too much focusing on a particular position or tokens has been prevented. On the basis of the context provided by the nonmasked words in the sequence, the model tries to predict the original value of the masked words.

These processes that need to be followed for generation of word embedding using BERT:

- Addition of a classification layer on top of the encoder output.

- Multiplication of the output vectors by the embedding matrix, thus transforming them into the vocabulary dimension.

- Calculation of the probability of each word in the vocabulary with softmax.

The loss function in BERT only considers the prediction of the masked values; the prediction of the nonmasked words is ignored. This makes the model converge slower than directional ones. As an example, for the sentence "The birds are flying in the clear blue sky," if we are training the bidirectional model instead of predicting the next word in the sequence, a model can be built to predict the missing word from within the sequence itself. Now, consider a token "flying" and mask it so that it can be considered missing. The model would now need to be trained in such a way that it can predict the value of this missing or masked token in the sentence "The birds are [MASK] in the clear blue sky." This is the essence of MLM, which enables the model to understand the relationships between words in a sentence.

Next Sentence Prediction

The NSP task is similar to next word prediction in a sentence. NSP predicts the next sentence in document, whereas the latter works for prediction of missing words in a sentence. BERT is also trained on the NSP task. This is required so that our model is able to understand how different sentences in a text corpus are related to each other. During the training of the BERT model, the sentence pairs are taken as input. It then predicts if the second sentence in the pair is the subsequent sentence in the original document. To achieve this, 50% of inputs are taken such that the second sentence is the subsequent sentence as in the original document, whereas the other 50% comprises the pair where the second sentence is chosen randomly from the document. It is assumed that the random second sentence is disconnected from the first sentence.

As an example, consider two different instances of training data for Sentence A and Sentence B:

```
Instance 1
Sentence A - I saw a bird flying in the sky.
Sentence B - It was a blue sparrow.
Label - IsNextSentence
Instance 2
Sentence A - I saw a bird flying in the sky.
Sentence B - The dog is barking.
Label - NotNextSentence
```

As we can see, for Instance 1 Sentence B is logically subsequent to Sentence A, but the same is not true for Instance 2, which is quite clear from the labels IsNextSentence and NotNextSentence, respectively.

These inputs are being processed even before the training process starts to differentiate between two sentences. The procedure is outlined here.

1. Two tokens are inserted in a sentence pair. One of the tokens [CLS] is inserted at the beginning of the first sentence and other token [SEP] is inserted at the end of each sentence. The two sentences are both tokenized and separated from each other by the separation token and then fed as a single input sequence into the model.

2. For each of the token sentences, embedding is added that indicates whether it is Sentence A or Sentence B. These sentence embeddings are basically similar in concept to token embeddings with a vocabulary of 2.

3. Along with the sentence embeddings, positional embeddings are also added to each of the tokens, which helps to indicate the position of the token in the sequence.

Now, the following steps are performed to predict if the second sentence is actually connected to the first.

1. The whole input sequence is passed though the transformer model.

2. With the help of the simple classification layer, the output of the [CLS] token is transformed into a 2X1 shaped vector.

3. Thereby, the probability of IsNextSentence is computed with the help of softmax.

As we know, BERT is used for variety of NLP tasks such as document summarization, question answering systems, document or sentence classification, and so on. Now, let's see how BERT can be used for classification of sentences.

Text Classification Using BERT

BERT can be used for a variety of language tasks. A small layer added to the core model allows use of BERT for tasks like classification, question answering, named entity recognition, and so on. The BERT model is fine tuned for this purpose. For classification tasks, a classification layer is added on top of the transformer output for the [CLS] token, similar to NSP. Most of the hyperparameters stay the same as in BERT training, but some of them require tuning to achieve state-of-the art-results for text classification tasks. Figure 4-2 gives an example of determining whether a given tweet is hate speech or not.

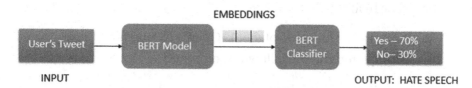

Figure 4-2. *An example of classification using BERT*

Similar types of tasks such as such as document classification, sentiment analytics, and so on, can also be achieved using BERT.

Next, we will see how a pretrained model of text classification can be configured in your system. Follow the steps listed here to configure or install the necessary prerequisites.

1. Make sure Python is installed on your system.
    Open a command prompt and run the following
    command to determine if Python is installed, as
    shown in Figure 4-3.

    Python

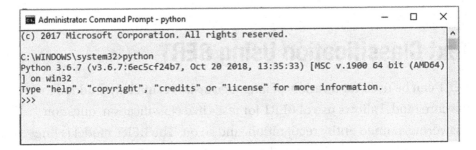

Figure 4-3. *Python console*

> This will start the Python console at the command
> prompt. If Python is not installed on your system,
> download and install Python as per your operating
> system using this link: https://www.python.org/
> downloads/

2. Next, install Jupyter Notebook. Open a command
    prompt and run the following command.

    pip install notebook

3. Open a command prompt and run the following
    command to run Jupyter Notebook.

    jupyter notebook

    The notebook will start in your default browser with
    localhost as the host address and port number as 8888,
    along with a unique token ID, as shown in Figure 4-4.

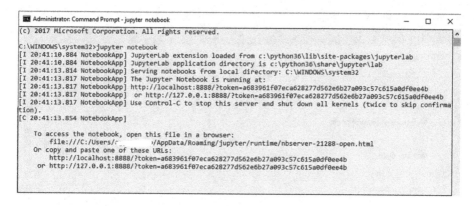

Figure 4-4. Jupyter Notebook console

4. You can also use Google Colab Notebook for
    the same purpose. It provides a fast and free
    environment to run your Python code in case your
    system doesn't have sufficient resources available.
    You can also use the graphics processing units
    (GPUs) and Tensor Processing Units (TPUs) for
    free, but for a limited time (12 hours) in Google
    Colab. You just need a Google account to log in to
    Google Colab Notebook. For this book, we will be
    using Google Colab Notebook to demonstrate text
    classification using BERT. Log in to your Google
    account and click https://colab.research.
    google.com. You will see the screen shown in
    Figure 4-5.

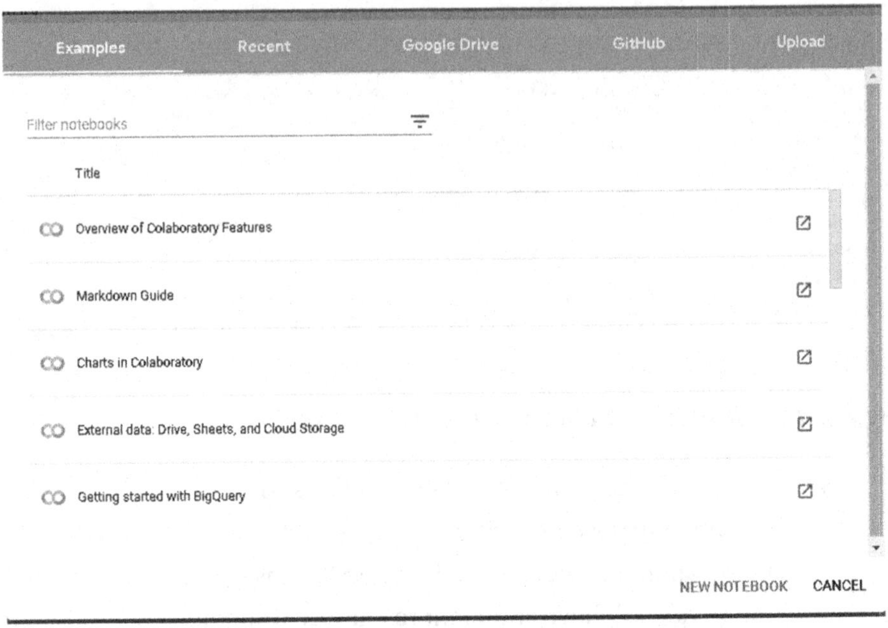

Figure 4-5. *Google Colab interface to create or import a notebook*

5. To create a new Colab notebook, click New
    Notebook in the bottom right corner as shown in
    Figure 4-5.

6. Install TensorFlow. Run the following command in
    your Jupyter Notebook or Colab Notebook.

```
pip install tensorflow
```

We have now installed all prerequisites for this exercise. Please follow
the steps listed next to configure a pretrained model for text classification
using BERT.

1. BERT model files and required code can be
    downloaded from the GitHub repository. Open the
    command prompt and clone the GitHub repository
    (i.e., `https:/github.com/google-research/`
    `bert.git`) onto the system by typing the following
    command:

    ```
    git clone https://github.com/google-research/bert.git
    ```

2. Downloaded model files containing the weights
    and other necessary BERT files. Depending on your
    requirements, a BERT pretrained model needs to be
    selected from this list.

    -   BERT Base, Uncased

    -   BERT Large, Uncased

    -   BERT Base, Cased

    -   BERT Large, Cased

3. If you have an access to a cloud TPU, BERT Large
    can be used; otherwise, the BERT Base model
    should be used. Further selection can be made from
    the cased and uncased models.

4. The data for fine-tuning the BERT model are
    expected to be in the format that BERT understands.
    Data have to be divided into three parts: train, dev,
    and test. As a rule of thumb, train should contain
    80% of data and the remaining 20% will be divided
    into dev and test. You need to make a folder
    containing three separate files: `train.tsv`, `dev.`
    `tsv`, and `test.tsv`. The `train.tsv` file will be used
    for training the model, `dev.tsv` will be used for

developing the system, and `test.csv` will be used for evaluating model performance over unseen data. Both `train.tsv` and `dev.tsv` should not have headers and should have four columns as shown below.

1	1	a	text example belonging to class 1
2	1	a	text example belonging to class 1
3	2	a	text example belonging to class 2
4	0	a	text example belonging to class 0

Here are details of the columns used.

a. **First Column**: Represents IDs of sample.

b. **Second Column**: Classification labels corresponding to examples.

c. **Third Column**: Throw-away column.

d. **Fourth Column**: This represents the actual textual sentence that needs to be classified.

5. The `test.tsv` file should have a header line, unlike the other two files, and should appear as shown here.

id sentence

1. first test example

2. second test example

3. third test example

6. To train the model, you need to navigate to the directory into which the model has been cloned. Afterward, enter the following command at the command prompt:

```
python bert/run_classifier.py \
--task_name=cola \
--do_train=true \
--do_eval=true \
--data_dir=./data \
--vocab_file=$BERT_BASE_DIR/vocab.txt \
--bert_config_file=$BERT_BASE_DIR/bert_config.json \
--init_checkpoint=$BERT_BASE_DIR/bert_model.ckpt \
--max_seq_length=128 \
--train_batch_size=32 \
--learning_rate=2e-5 \
--num_train_epochs=3.0 \
--output_dir=./bert_output/
```

If length of your training data text is longer than 128 words then the value for max_seq_length can be increased to 512. If you are training the model over a CPU system, then you can reduce the training size to avoid an out-of-memory error.

When the training is finished, the reports get stored in the bert_output directory.

7. This trained BERT model is now ready to use for prediction purposes. If we have to make a prediction for new data, then data need to be stored in test. tsv. Go to the directory where the trained model files have been stored. Please refer to the highest-

number (latest model file) model.ckpt file seen in the bert_output directory. These files contain the weights of the model trained. Now run the following commands at the command prompt to obtain the classification result, which will be stored in test_results.tsv in the bert_output directory location.

```
python bert/run_classifier.py \
--task_name=cola \
--do_predict=true \
--data_dir=./data \
--vocab_file=$BERT_BASE_DIR/vocab.txt \
--bert_config_file=$BERT_BASE_DIR/bert_config.json \
--init_checkpoint=$TRAINED_CLASSIFIER \
--max_seq_length=128 \
--output_dir=./bert_output/
```

Please note that the value for the max_seq_length parameters should be the same as what was used during the training process.

For this book, we will demonstrate implementation of a question classification dataset where questions will be classified into their respective categories. There are mainly two types of questions, factoid (nondescriptive) and non-factoid questions. As an example, "What is the temperature in Delhi?" is a factoid question, as it is looking for an answer based on some facts. "What is temperature?" is a non-factoid question, as it is looking for text snippets about temperature. For this implementation, please refer to the dataset at https://cogcomp.seas.upenn.edu/Data/QA/QC/.

Now we will see how a question classification system can be implemented using BERT.

1. For this implementation, we will download the
    BERT base-cased model from GitHub as described
    previously.

2. The question classification dataset is already in the
    format required for training the BERT model. The
    data are split into train.tsv, dev.tsv, and test.
    tsv sets. In train.tsv and dev.tsv, we do not have
    any headers. The following is a description of the
    columns in the file.

    •   **First Column**: Index for data point.

    •   **Second Column**: Classification label (i.e., factoid or
        non-factoid). In this dataset, factoid is represented
        by 0 and 1 for non-factoid.

    •   **Third Column**: Throwaway column with value a.

    •   **Fourth Column**: Actual question text.

    Then we create data folder and save these files in the
    folder. Please refer Figures 4-6 through 4-8 for some
    examples from training files.

```
1    0   1   a   How did serfdom develop in and then leave Russia ?
2    1   0   a   What films featured the character Popeye Doyle ?
3    2   1   a   How can I find a list of celebrities' real names ?
4    3   0   a   What fowl grabs the spotlight after the Chinese Year of the Monkey ?
5    4   0   a   What is the full form of .com ?
6    5   0   a   What contemptible scoundrel stole the cork from my lunch ?
7    6   0   a   What team did baseball's St. Louis Browns become ?
8    7   0   a   What is the oldest profession ?
9    8   1   a   What are liver enzymes ?
```

Figure 4-6. *Snapshot of Dev.tsv*

```
1       55  1   a   What is the nature of learning ?
2       56  0   a   What's the only color Johnny Cash wears on stage ?
3       57  0   a   What's the term for a young fox ?
4       58  0   a   What U.S. state lived under six flags ?
5       59  1   a   What is the Kashmir issue ?
6       60  0   a   Where is the Loop ?
```

Figure 4-7. *Snapshot of* `train.tsv`

```
1       User_ID Description
2       1   I have a mac OS and I want to upgrade it to latest. Please help
3       2   My OS X is 10.8 , how to upgrade it ?
4       3   What is the process to update Mac OS?
5       4   I am no more able to see my deleted email, can you help?
6       5   I accidently deleted my emails, is there a way I can i recover them
```

Figure 4-8. *Snapshot of* `test.tsv`

3. Now, navigate to the directory where the downloaded BERT model exists.

4. As mentioned earlier, execute the command for training at the command prompt. The model output after completion of training gets stored in the location that has been defined under the bert_ output parameter, as shown in Figure 4-9.

```
python run_classifier.py --task_name=cola --do_
train=true --do_eval=true --data_dir=$BERT_
BASE_DIR/data --vocab_file=$BERT_BASE_DIR/
bert_output/cased_L-12_H-768_A-12/vocab.
txt --bert_config_file=$BERT_BASE_DIR/bert_
output/ cased_L-12_H-768_A-12/bert_config.
json --init_checkpoint=$BERT_BASE_DIR/bert_
output/ model.ckpt-2023 --max_seq_length=
128 --train_batch_size=32 --learning_
rate=2e-5 --num_train_epochs=3.0 --output_
dir=$BERT_BASE_DIR/bert_output/
```

$BERT_BASE_DIR is a directory where you must have
downloaded code from GitHub.

```
C:\BERT_base\bert>python run_classifier.py --task_name=cola --do_train=true --do_eval=true --data_dir=C:/BERT_base/base/
data --vocab_file=C:/BERT_base/base/bert_output/cased_L-12_H-768_A-12/vocab.txt --bert_config_file=C:/BERT_base/base/ber
t_output/ cased_L-12_H-768_A-12/bert_config.json --init_checkpoint=C:/BERT_base/base/bert_output/ model.ckpt-2023 --max_
seq_length=128 --train_batch_size=32 --learning_rate=2e-5 --num_train_epochs=3.0 --output_dir=C:/BERT_base/bert_out
put/
```

Figure 4-9. *Command to train BERT model*

5. After completion of training, we can classify the test
    data using the trained model. Run the following
    command at the command prompt to get a
    prediction for questions present in the test.tsv file
    as shown in Figure 4-10.

    python bert/run_classifier.py --task_name=cola --do_
    predict=true --data_dir=$BERT_BASE_DIR/data
    vocab_file=$BERT_BASE_DIR/bert_output/cased_L-
    12_H-768_A-12/vocab.txt --bert_config_file=$BERT_
    BASE_DIR/bert_output/ cased_L-12_H-768_A-12/
    bert_config.json --init_checkpoint=$TRAINED_
    CLASSIFIER --max_seq_length=128 --output_
    dir=$BERT_BASE_DIR/bert_output/

    $BERT_BASE_DIR is a directory where you must have
    downloaded code from GitHub.

```
C:\BERT_base\bert>python bert/run_classifier.py --task_name=cola --do_predict=true --data_dir=C:/BERT_base/base/data voc
ab_file=C:/BERT_base/base/bert_output/cased_L-12_H-768_A-12/vocab.txt --bert_config_file=C:/BERT_base/base/bert_output/
cased_L-12_H-768_A-12/bert_config.json --init_checkpoint=$TRAINED_CLASSIFIER --max_seq_length=128 --output_dir=C:/BERT_b
ase/base/bert_output/
```

Figure 4-10. *Command for prediction*

6. The results of the classification are stored in the location that has been defined as the value for the bert_output parameter in the test_results. tsv file. As we can see in Figure 4-11, the result of classification is a probability distribution of a question to two classes. The class with the higher score will be considered the relevant one.

0.9989654	0.0010346028
0.9989654	0.0010346028
0.9989654	0.0010346028
0.99344105	0.006558984
0.9989654	0.0010346028
0.9989654	0.0010346028
0.9989654	0.0010346028
0.9989654	0.0010346028
0.99344105	0.006558984
0.99344105	0.006558984
0.9989654	0.0010346028
0.9989654	0.0010346028
0.9989654	0.0010346028
0.9989654	0.0010346028

Figure 4-11. *Prediction results snapshot*

The first column corresponds to the label 0 (Factoid) and the second column corresponds to the label 1 (non-factoid). From this generated .csv we can see whether the questions in the test data are Factoid or non-factoid questions.

This question type classification system is quite useful in a conversational system where a query or question entered by an end user needs to be classified to retrieve relevant results.

Benchmarks for BERT Model

BERT embedding model performance and accuracy have been continuously evaluated over different types of datasets for various NLP tasks. This is being done to check if BERT is able to achieve benchmark values already set up for these datasets by some other methods. These benchmarks are datasets that evaluate the working of specific aspects of a model. There exist many such benchmarks and some of them are discussed next.

GLUE Benchmark

General Language Understanding Evaluation (GLUE) is a collection of datasets that can be used to train, evaluate, and analyze NLP models. These different models are compared with each other over the GLUE dataset. To test a model's language understanding, the GLUE benchmark includes nine diverse task datasets. To evaluate a model, first it is trained over a dataset provided by GLUE and then it is scored on all nine tasks. The final performance score is the average of all nine tasks.

$$Final\ GLUE\ Score = \sum Individual\ Task\ Score$$

The model is required to have representation of its input and output changed so as to accommodate the task. For instance, during the pretraining of BERT, few words are masked when sentences are given as input. Because the input representation layer in BERT accommodates all of the GLUE tasks, there is no need to change this layer. However, the pretraining classification layer has to be removed. This layer is replaced with the one that accommodates the output of each GLUE task. The BERT model scores a state-of-the-art result on the GLUE benchmark, with a score of 80.5%.

SQuAD Dataset

The Stanford Question Answering Dataset (SQuAD) is a reading comprehension dataset, consisting of questions asked on a set of Wikipedia articles. The answer to each of the questions is either a text segment or a span from the passage, respectively. There are two versions of the SQuAD dataset.

- SQuAD 1.1

- SQuAD 2.0

SQuAD2.0 has 100,000 questions in addition to SQuAD 1.1, which contains 50,000 unanswered questions, but are similar to questions that were answerable. This was done so that SQuAD2.0 can do well in situations where no answers to questions are supported by a paragraph for a question.

BERT is able to achieve state-of-the-art results on the SQuAD dataset with minor modifications. It requires semicomplex preprocessing of data and postprocessing to deal with the variable-length nature of SQuAD context paragraphs and the character-level answer annotations used for SQuAD training. The BERT model was able to achieve an F1 score of 93.2 and 83.1 for SQuAD 1.0 and SQuAD v2.0 over test dataset, respectively.

IMDB Reviews Dataset

The IMDB dataset is an extensive movie review dataset that has been used for classification of viewer sentiments about films. This dataset consists of 25,000 highly polar movie reviews for training and 25,000 reviews for testing. In addition to the training and testing data, there are also additional unlabeled data. This dataset has also been used to evaluate BERT in a sentiment classification task.

RACE Benchmark

RACE is a large-scale reading comprehension dataset from examinations. The RACE dataset is used to evaluate models on a reading comprehension task. This dataset was collected from English examinations of Chinese students. It consists of nearly 28,000 passages and 100,000 questions generated by human experts. The number of questions is much larger in RACE as compared to other benchmark datasets. The BERT large model achieves a score of 73.8% on the RACE benchmark dataset.

Types of BERT-Based Models

BERT is a ground-breaking natural language model and its introduction in the ML world has led to development of various models that are based on it. Variants of the BERT model have been developed to cater to different types of NLP-based systems. Here are a few of the major variants of BERT:

- ALBERT

- RoBERTa

- DistilBERT

- StructBERT

- $BERT_{joint}$ for Natural Questions

ALBERT

ALBERT is a much smaller version of BERT that was introduced jointly by Google Research and the Toyota Technological Institute. It is a smarter, "lite" BERT and is also considered a natural successor to BERT. It can also be used to implement state-of-the-art NLP tasks. This is all possible with less computation power compared to BERT, but you need to compromise on accuracy a little bit. ALBERT was basically created to make improvements in architecture and training methods so that better results are delivered with fewer required computation resources.

ALBERT has a BERT-like core architecture. It has a transformer encoder architecture and a vocabulary of 30,000 words, which is the same as BERT. However, there are substantial architectural improvements in ALBERT for efficient parameter usage.

1. **Factorized embedding parameterization:** In the case of BERT, the WordPiece embeddings size (E) is directly tied to the hidden layer size (H). It was observed that WordPiece embeddings are designed to learn context-independent representations, whereas the hidden layer embeddings are designed to learn context-dependent representations. In BERT we try to learn context-dependent representations through the hidden layers only.

   When H and E are tied together, we end up with a model with billions of parameters that are rarely updated in training. This happens as the embedding matrix, which is V*E where V is the large vocabulary, must scale with the H (hidden layers). This actually results in inefficient parameters, as these two items work for different purposes.

   In ALBERT, to make it more efficient we untie the two parameters and embedding parameters are split into two smaller matrices. Now the one-hot vectors are not directly projected into H; rather, they are projected into a smaller, lower dimension matrix E, and then E is projected into the hidden layers. Thus, the parameters get reduced from O (V*H) to $\Theta(V*E+E*H)$.

2. **Cross-layer parameter sharing:** ALBERT has a smoother transition from layer to layer in comparison to BERT and the parameter efficiency is improved by sharing of all the parameters across all layers. The feed-forward and attention parameters are all shared. This weight sharing is helpful in stabilizing the network parameters.

3. **Training changes: Sentence order prediction:** Similar to BERT, ALBERT also uses MLM but does not use NSP. Instead of NSP, ALBERT uses its own newly developed training method called sentence order prediction (SOP).

   The NSP loss used in BERT was not found to be a very effective training mechanism in subsequent studies. Hence, it was leveraged to develop SOP as NSP was unreliable.

   In ALBERT SOP, loss is used to model intersentence coherence. SOP was mainly created to focus on intersentence coherence loss instead of topic prediction, whereas BERT combines topic prediction with coherence prediction. Hence, ALBERT is able to learn finer grained intersentence cohesion by avoiding issues of topic prediction.

ALBERT, even though it has fewer parameters than BERT, gets results in less time. In the language benchmark tests SQuAD1.1, SQuAD2.0, MNLI SST-2, and RACE, ALBERT has significantly outperformed BERT, as we can see in the comparison in Table 4-1.

Table 4-1. *Comparison Between BERT and ALBERT Models*

Model	Parameters	SQuAD1.1	SQuAD2.0	MNLI	SST-2	RACE	Avg	Speedup
BERT base	108M	90.5/83.3	80.3/77.3	84.1	91.7	68.3	82.1	17.7x
BERT large	334M	92.4/85.8	83.9/80.8	85.8	92.2	73.8	85.1	3.8x
BERT xlarge	1270M	86.3/77.9	73.8/70.5	80.5	87.8	39.7	76.7	1.0
ALBERT base	12M	89.3/82.1	79.1/76.1	81.9	89.4	63.5	80.1	21.1x
ALBERT large	18M	90.9/84.1	82.1/79.0	83.8	90.6	68.4	82.4	6.5x
ALBERT xlarge	59M	93.0/86.5	85.9/83.1	85.4	91.9	73.9	85.5	2.4x
ALBERT xxlarge	233M	94.1/88.3	88.1/85.1	88.0	95.2	82.3	88.7	1.2x

RoBERTa

RoBERTa is an optimized method for pretraining NLP systems. RoBERTa (Robustly optimized BERT) was developed by the Facebook AI team and based on Google's BERT model. RoBERTa reimplemented the neural network architecture of BERT with additional pretraining improvements that achieve state-of-the-art results on several benchmarks.

RoBERTa and BERT share several configurations, but there are some model settings that differentiate the two models.

- **Reserved token:** BERT uses [CLS] and [SEP] as starting token and separator token, respectively, whereas RoBERTa uses <s> and </s> to convert sentences.

- **Size of sub-word:** BERT has about 30,000 sub-words, whereas in RoBERTa there are about 50,000 sub-words.

In addition, there are specific modifications and adjustments that help RoBERTa to perform better than BERT.

- **More training data:** During reimplementation of BERT, several changes were made to the hyperparameters of the BERT model and training was done with a higher magnitude of data with more iterations. RoBERTa uses more training data. It uses BookCorpus (16G), CC-NEWS (76G), OpenWebText (38G), and Stories (31G) data, whereas BERT uses only BookCorpus as training data.

- **Dynamic masking:** When BERT was being ported to create RoBERTa, the creators modified the word masking strategy. BERT mainly uses static masking, in which the words are masked from sentences during preprocessing. RoBERTa makes use of dynamic masking. Here, a new masking pattern is generated whenever a sentence is fed into training. RoBERTa duplicates training data 10 times and masks those data differently. It is experimentally observed that the dynamic masking improves performance and gives a better result than static masking.

- **Different training objective:** BERT captures the relationships between the sentences by training on NSP. Some training approaches without application of NSP provided better results, proving the ineffectiveness of NSP. Experiments were done to compare models trained with segment-pair with NSP, sentence-pair with NSP, full sentences without NSP, and doc-sentences without NSP. The models trained without NSP performed better on SQuAD1.1/2.0, MNLI-m, SST-2, and RACE.

- **Training on longer sequences:** Better results have
  been achieved when a model was trained on longer
  sequences. BERT base is trained with a batch size of
  256 sequences via 1 million steps, but training on 2,000
  sequences and 31,000 steps shows improvement in
  performance.

With the implementation of the design changes, the RoBERTa model
delivered state-of-the-art performance on the MNLI, QNLI, RTE, and
RACE tasks. It also realized a sizable performance improvement on the
GLUE benchmark with a score of 88.5.

RoBERTa demonstrates that the tuning the BERT training procedure
can result in performance improvement on a variety of NLP tasks. This
highlights the importance of exploring the design choices in BERT training
for better performance output.

DistilBERT

DistilBERT was introduced for knowledge distillation. This knowledge
distillation was required to address the drawbacks of computation of
large numbers of parameters. The NLP models that have been developed
recently show an increase in parameter count, now reaching parameter
counts as high as in the tens of billions. Even though higher parameter
count ensures optimal performance, it prevents model training and
serving when computational resources are limited.

Knowledge distillation revolves around the idea that a larger model
acts as a teacher for a smaller one that tries to replicate its outputs and
sublayer activation for a given set of inputs. This is sometimes also known
as teacher–student learning. It is a compression technique where the
behavior of larger models is reproduced by the smaller ones. The output
distribution from the teacher can be used for all possible targets, which
helps in creation of a student with generalizability. For example, in the

sentence "The sky is [mask]" a teacher might assign high probabilities to words like "cloudy" and "clear." There are also chances that a high probability is assigned to the word "blue." This is helpful for the student model, so that it is able to generalize rather than only learn the correct target. This information is captured through the loss function that is being used to train the student. This loss function comprises a linear combination of three factors.

Distillation Loss

Distillation loss takes into consideration combination of the output probabilities of the teacher (t) and the student (s) as shown in the following equation.

$$L_{ce} = \sum_i t_i \log(s_i)$$

Distillation Loss

$$t_i = \exp(z_i/T)/ \sum_j \exp(z_j/T)$$

Temperature Softmax

The teacher probabilities are calculated through temperature softmax. This is basically a modification to the softmax so that more granularities are obtained from the teacher model output distribution. This gives a smoother output distribution, as the size of larger probabilities are decreased and the smaller ones are increased. This helps to minimize the distillation loss.

Cosine Embedding Loss

Cosine embedding loss is a distance measure between the hidden representations for teacher and student. This helps in creation of a better model as it enables the student to imitate the teacher not just in the output layer, but in other layers, too.

Masked Language Modeling Loss

This is the same loss as used in training of the BERT model to predict the correct token value for the masked token in the sequence.

Architectural Modifications

The DistilBERT network architecture is also a transformer encoder model similar to BERT base, but it has half the number of layers. The hidden representations, though, are kept the same. This affects the parameter count, with a 66 million parameter count in the case of DistilBERT, whereas there are 110 million parameters in the teacher model. The reduction in the model size through the number of layers helps to achieve the drastic reduction in computation complexity. The reduction in the size of the vectors or the hidden state representations have also reduced the model size.

After the knowledge distillation, DistilBERT is able to achieve 97% of BERT base's score on the GLUE benchmark. This knowledge distillation has helped to condense the larger models or ensembles of models into a smaller student network. This has proven to be helpful in situations where the computational environment is limited.

StructBERT

StructBERT is a model based on BERT that incorporates language structures into BERT pretraining. The two linearization strategies help to incorporate language structure into BERT. Word-level ordering and sentence-level ordering are the two structural information sets that are leveraged in StructBERT. StructBERT achieves better generalizability and adaptability due to the incorporation of this structural pretraining. The dependency between the words as well as sentences is encoded in StructBERT.

Structural Pretraining in StructBERT

Similar to all the other BERT-based models, StructBERT also builds on the BERT architecture. The original BERT performs two unsupervised pretraining tasks, MLM and NSP. StructBERT is able to increase the ability of the MLM task. It shuffles a certain number of tokens after masking of words and predicts the right order. StructBERT is also able to understand the relationship between sentences in a better way. This is achieved by random swapping of the sentence order. This new BERT-based model captures the fine-grained word structure in every sentence.

After pretraining of the StructBERT it can be fine-tuned on task-specific data for a wide range of downstream tasks such as document summarization.

Pretraining Objectives

The pretraining objectives of the original BERT are extended in the case of StructBERT to fully utilize rich inner-sentence and intersentence structures in language. This is done in two ways.

1. **Word structural objective:** The BERT model fails to model sequential order and high-order word dependency in natural language explicitly. A good language model should be able to reconstruct a sentence from a given sentence that has randomly ordered words. StructBERT is able to implement this idea by supplementing BERT's training objectives with a new word structural objective. This new model objective gives the model the ability to restructure the sentence to have correct ordering of the randomly shuffled word tokens. This objective is trained together with the MLM objective from BERT.

2. **Sentence structural objective:** The original BERT model objective of NSP is extended in StructBERT by predicting both the next sentence as well as the previous sentence. This makes the pretrained model learn the sequential ordering of the sentences in a bidirectional manner.

These two auxiliary objectives are pretrained together with the original MLM objective to exploit inherent language structures.

BERT$_{joint}$ for Natural Questions

BERT$_{joint}$ is a BERT-based model for Natural Questions. The BERT$_{joint}$ model predicts short and long answers in a single model only instead of a pipeline approach. In this model each document is split into multiple instances of training with the help of overlapping token windows. This approach is used to create a balanced training set and is being followed by down sampling instances without an answer (null instances). The [CLS] token is used during training to predict null instances, and spans at inference time are ranked by the difference between the span score and the [CLS] score.

The model uses the Natural Questions (NQ) dataset that is a question answering dataset of 307,373 training examples, 7,830 development examples, and 7,842 test examples. For every example, a query is entered by the user over the Google search engine and the corresponding Wikipedia page that contains an answer. The Wikipedia page is annotated as an answer to the question.

The BERT$_{joint}$ model was initialized from the original BERT model that trained on the SQuAD 1.1 dataset. Afterward, this model was fine-tuned on Natural Questions training instances. It has used the Adam optimizer to minimize the loss. The BERT$_{joint}$ model for Natural Questions gives dramatically better results than the baseline NQ systems. This variation of BERT offers a new way to design a question-answering system.

Conclusion

This chapter looked deeper into BERT, along with its two algorithms, MLM and NSP. We also discussed a sample text classification model developed using BERT. We also examined the behavior of BERT over different benchmark datasets, along with multiple variations of BERT. In the next chapter, we discuss the design of a question answering system using BERT.

CHAPTER 5

BERT Model Applications: Question Answering System

We are surrounded by massive amounts of information present in the form of documents, images, blogs, websites, and more. In most cases, we always look for a direct answer instead of reading the entirety of lengthy documents. Question answering systems are generally being used for this purpose. These systems scan through a corpus of documents and provide you with the relevant answer or paragraph. It is part of the computer science discipline in the field of information retrieval and NLP, which focuses on building systems that automatically extract an answer to questions posed by humans or machines in a natural language.

Two of the earliest question answering systems, BASEBALL and LUNAR, have been popular because of their core database or information system. BASEBALL was built for answers to American League baseball

© Navin Sabharwal, Amit Agrawal 2021
N. Sabharwal and A. Agrawal, *Hands-on Question Answering Systems with BERT*,
https://doi.org/10.1007/978-1-4842-6664-9_5

questions over a one-year cycle. LUNAR was built to answer questions related to geological analysis of lunar rocks based on data collected from the Apollo moon mission. Such earlier systems concentrated on closed domains where every query must be about the specific domain and the answer text must be from a restricted vocabulary only.

Some of the advanced question answering systems of the modern world are Apple Siri, Amazon Alexa, and Google Assistant. There are various popular datasets available for question answering systems that can be leveraged to check your model performance. These include the following.

- **SQuAD**: The Stanford Question Answering Dataset (SQuAD) is a reading comprehension dataset that we covered in Chapter 4.

- **NewsQA**: This dataset has been created to help the research community build algorithms that are capable of answering questions requiring human-level comprehension and reasoning skills. By using CNN articles from the DeepMind Q&A dataset, authors have prepared a crowd-sourced machine reading comprehension dataset of 120,000 Q&A pairs.

- **WikiQA**: This publicly available dataset contains pairs of questions and answers. It has been collected and annotated for research on open-domain question answering systems. In addition, the WikiQA dataset also includes questions for which there are no correct answers, enabling researchers to work on negative cases as well to avoid selection of irrelevant answers.

Types of QA Systems

Question answering systems are broadly divided into two categories: open-domain QA (ODQA) system and closed-domain QA (CDQA) system.

- **Closed-domain:** In closed-domain systems, questions belong to a particular domain. They can answer the questions from a single domain only. As an example, a question answering system for the health care domain cannot answer any IT-related questions. These systems exploit domain-specific knowledge by using a model that has been trained on a domain specific dataset. The CDQA suite can be used to build such a closed-domain QA system.

- **Open-domain:** In open-domain systems, questions can be from any domain, such as health care, IT, sports, and more. These systems are designed to answer questions from any domain. These systems actually mimic human intelligence to answer questions. One example of such a system is the DeepPavlov ODQA system, an ODQA developed by MIPT that uses a large dataset of articles from Wikipedia as its source of knowledge.

These systems can be further divided into factoid and non-factoid as briefly covered in Chapter 4 and described here.

- **Factoid question:** A factoid question is about providing concise facts. Answers to factoid questions are based on proven facts. As an example, a learner might be asked to look at a passage, then answer a series of factual questions based on what he or she just read. These types of questions usually start with who, what, when, or where.

Here are some examples of factoid questions.

- Who is the president of the United States?

- Who is the prime minister of India?

- Who is the CEO of Google?

All of these questions can be answered from any document or blog if text contains relevant data which is sufficient to answer questions.

- **Non-factoid question**: A non-factoid question expects detailed answers about any topic. As an example, a user can ask questions related to mathematical problems, how to run a vehicle, what does temperature mean, and so on. Non-factoid questions usually require multiple sentences as answers, and these answers come from a particular paragraph in a document. Thus, the context of a sentence plays an important role to retrieve the relevant answer.

Here are some examples of non-factoid questions.

- What is the process of installing Python on Windows?

- How can I reset my Microsoft Outlook password?

- What do you mean by temperature?

Answers to these questions will be a document, a paragraph, or a definition from a paragraph.

Question Answering System Design Using BERT

This section details how BERT can be used for implementation of a factoid question answering system. For this book, we are using a pretrained model that has been trained on the SQuAD version 1 dataset.

As an example, consider this question, along with the passage from a Wikipedia article on the Football League.

> **Question**: Where was the Football League founded?

> **Passage:** In 1888, The Football League was founded in England, becoming the first of many professional football competitions. During the 20th century, several of the various kinds of football grew to become some of the most popular team sports in the world.

The answer to this question will therefore be England.

Now, we look closer at how this question and passage will be processed using BERT to find the relevant answer. This is all in the context of a question answering system, compared to the text classification approach in Chapter 4.

BERT extracts tokens from the question and passage and combines them together as an input. As mentioned earlier, it starts with a [CLS] token that indicates the start of a sentence and uses an [SEP] separator to separate the question and passage. Along with the [SEP] token, BERT also uses segment embeddings to differentiate between the question and the passage that contains an answer. BERT creates two segment embeddings, one for the question and other for the passage, to differentiate between question and passage. Then these embeddings are added to a one-hot representation of tokens to segregate between question and passage as shown in Figure 5-1.

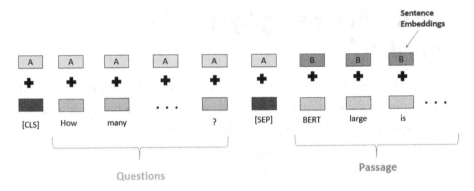

Figure 5-1. *BERT input representation*

Next, we pass our combined embedded representation of question and passage as input in the BERT model. The last hidden layer of BERT will then be changed and uses softmax to generate probability distributions for the start and end index over an input text sentence that defines a substring, which is an answer, as shown in Figure 5-2.

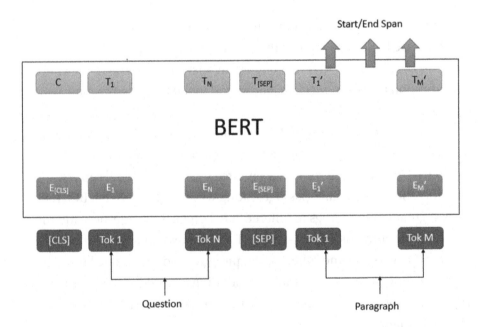

Figure 5-2. *BERT architecture for question answering system*

To this point, we have discussed how BERT will process input questions and passages. Next, we will see an implementation of a question answering system using BERT in Python.

Follow the steps given here to install the required prerequisites for a BERT-based question answering system. Many of them are the same as those for the examples in Chapter 4, but are included for completeness to ensure you can run the examples in this chapter.

1. Make sure Python is installed on your system. Open a command prompt and run the following command to determine if Python is installed, as shown in Figure 5-3.

   python

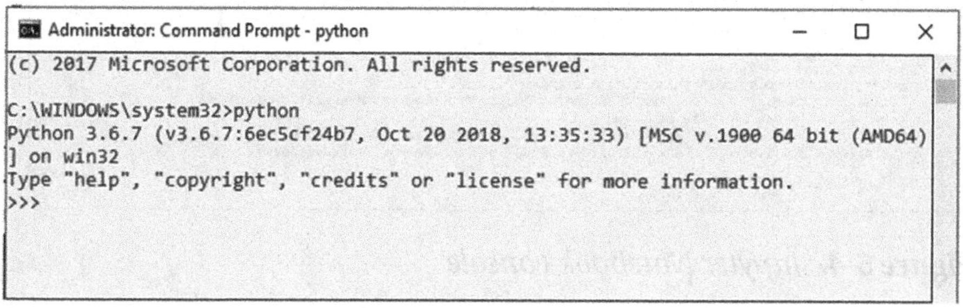

***Figure 5-3.** Python console*

This will open your Python console at the command prompt. If Python is not installed on your system, download and install it as per your operating system from https://www.python.org/downloads/.

2. Next, install Jupyter Notebook, which we will use to code. Open a command prompt and run the following command.

   pip install notebook

3. Open a command prompt and run the following command to run Jupyter Notebook.

```
jupyter notebook
```

The notebook will open in your default browser with the host address as localhost and the port number as 8888, along with a unique token ID. Now, you can start writing code as mentioned in subsequent steps, as shown in Figure 5-4.

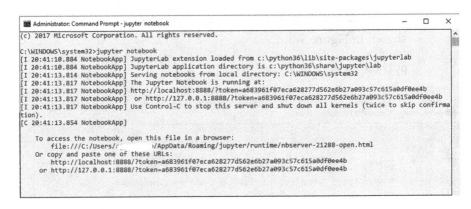

Figure 5-4. Jupyter Notebook console

4. You can also use Google Colab Notebook for the same purpose. It provides a fast and free environment to run your Python code if your system doesn't have sufficient resources available. You can also use the GPUs and TPUs for free but for a limited time (12 hours) in Google Colab. You just need a Google account to log in to Google Colab Notebook. For this book, we will be using Google Colab Notebook to demonstrate a question answering

system using BERT. Log in to your Google account and click https://colab.research.google.com. You will see the screen shown in Figure 5-5.

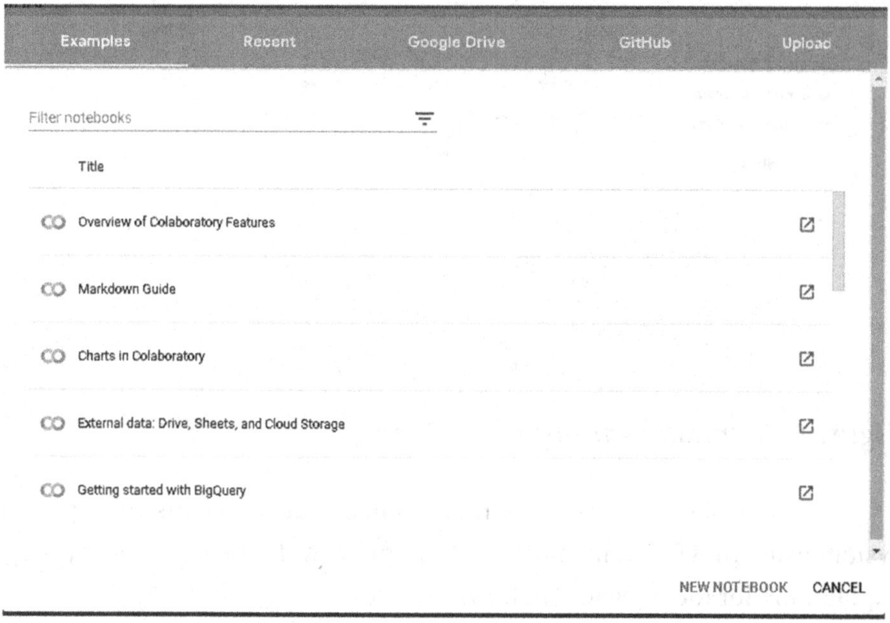

Figure 5-5. *Google Colab interface to create or import a notebook*

5. To create a new Colab notebook, click New Notebook in the bottom right corner as shown in Figure 5-5.

6. Install the transformers library from Huggingface. Run the following command in your Jupyter Notebook or Colab Notebook.

```
pip install transformers
```

7. After successful installation of the transformers library, you should be able to see the output shown in Figure 5-6.

Figure 5-6. *Installing transformers library*

Next, we will proceed to implementation of a question answering system using BERT. The included code snippets will provide a step-by-step explanation for the question answering system.

1. Import the `BertQuestionAnswering` and `BertTokenizer` classes of the transformers library as shown here.

```
from transformers import BertForQuestionAnswering
from transformers import BertTokenizer
import torch
```

2. Next, load the BERT question answering model fine-tuned on the SQuAD version 2 dataset. It will be a large version of BERT, with 24 layers, 340 million parameters, and an embedding size of 1,024. Along with the BERT model, we have also downloaded a trained model vocabulary set as shown here.

```
# Load pretrained model for Question
Answering
bert_model = BertForQuestionAnswering.from_
pretrained('bert-large-uncased-whole-word-
masking-finetuned-squad')

#Load Vocabulary
bert_tokenizer = BertTokenizer.from_
pretrained('bert-large-uncased-whole-word-
masking-finetuned-squad')
```

Note   This will take a few minutes, depending on your Internet bandwidth, as the model size is approximately 1.34 GB.

---

3. Next, it requires a question and candidate paragraph context where an answer to the question would exist. You can find a candidate paragraph using any search engine or document indexer system such as Apache Solr or Watson Discovery Service (WDS). These systems will provide context paragraphs for the question asked by the user.

4. Then, the question, along with the context paragraph, will be passed to the question answering system, where first they will be tokenized based on downloaded vocabulary. As mentioned earlier, these will be concatenated together using a special character [SEP] token in between as shown here (reference text has been taken from a Wikipedia article).

```
question = "Where was the Football League
founded?"
reference_text = " In 1888, The Football League
was founded in England, becoming the first of
many professional football competitions. During
the 20th century, several of the various kinds
of football grew to become some of the most
popular team sports in the world."

#Perform tokenization on input text
input_ids = bert_tokenizer.encode(question,
reference_text)
input_tokens = bert_tokenizer.convert_ids_to_
tokens(input_ids)
```

5. Next, we need to concatenate them using segment
    embedding to differentiate between the question
    and the context passage. Segment embedding for
    the question will be added to the token vector of
    the question and similarly for segment embedding
    for the context passage. This should be done before
    even using it as an input to the BERT model. These
    additions are managed internally by the transformer
    library, but we need to provide Boolean values (0 or 1)
    to differentiate for each token as shown here.

```
#Find index of first occurrence of [SEP] token
sep_location = input_ids.index(bert_tokenizer.sep_
token_id)
first_seg_len, second_seg_len = sep_location+1,
len(input_ids)-(sep_location+1)
seg_embedding = [0]*first_seg_len + [1]*second_seg_len
```

6. Now we can pass our example to the model.

```
#Test model on our example
model_scores=bert_model(torch.tensor([input_ids]),
token_type_ids=torch.tensor([seg_embedding]))

ans_start_loc, ans_end_loc = torch.
argmax(model_scores[0]),
torch.argmax(model_scores[1])

result = ' '.join(input_tokens[ans_start_
loc:ans_end_loc+1])
result = result.replace(' ##','')
```

7. The model will provide start and end index from
    context passage as an answer such as start index
    value as 11 and end index value as 18. The final
    output will be extracted from context passage using
    these indexes.

Here is the complete Python code that takes the question and
reference passage as an input and finds the answer to that question.

```
from transformers import BertForQuestionAnswering
from transformers import BertTokenizer
import torch

def get_answer_using_bert(question, reference_text):
    # Load pretrained model for Question Answering
    bert_model = BertForQuestionAnswering.from_
pretrained('bert-large-uncased-whole-word-masking-
finetuned-squad')

    #Load Vocabulary
    bert_tokenizer = BertTokenizer.from_pretrained('bert-large-
uncased-whole-word-masking-finetuned-squad')
```

```
#Perform tokenization on input text

input_ids = bert_tokenizer.encode(question, reference_text)
input_tokens = bert_tokenizer.convert_ids_to_tokens(input_
ids)
```

```
#Find index of first occurrence of [SEP] token
    sep_location = input_ids.index(bert_tokenizer.sep_token_id)
    first_seg_len, second_seg_len = sep_location+1, len(input_
    ids)-(sep_location+1)
    seg_embedding = [0]*first_seg_len + [1]*second_seg_len

    #Test model on our example
    model_scores = bert_model(torch.tensor([input_ids]), token_
    type_ids=torch.tensor([seg_embedding]))
    ans_start_loc, ans_end_loc = torch.argmax(model_scores[0]),
    torch.argmax(model_scores[1])
    result = ' '.join(input_tokens[ans_start_loc:ans_end_
    loc+1])

    result = result.replace(' ##','')
    return result

if __name__ == "__main__" :
question = "Where was the Football League founded?"
reference_text = " In 1888, The Football League was founded
in England, becoming the first of many professional football
competitions. During the 20th century, several of the various
kinds of football grew to become some of the most popular team
sports in the world."
print(get_answer_using_bert(question, reference_text))
```

After running this code in Colab Notebook, we get following output:

england

Now, we have seen how a BERT-based question answering system can be used for research purposes. Next, consider a scenario where you need to deploy this feature to be consumed by some website or conversation system to serve the end user who is looking for an answer to his or her query. In this case, you need to release or expose features of the QA system as a REST API. Now, follow below steps to release features of QA system as REST API.

Let's go through the steps to set up a REST API and public URL for that API (use ngrok to generate a public URL if you are inside the private network) for a question answering system on both Windows and Linux Server.

For Windows Server

Prerequisite: Python 3.6.x and Pip need to be installed on your system.

Creation of REST API

1. Install Flask-RESTful

Flask-RESTful is an extension of the micro-framework Flask for building REST APIs.

For installation, run the following command at the Windows command prompt, as shown in Figure 5-7.

```
pip install flask-restful
```

```
C:\Windows\System32\cmd.exe                                                    —    □    ×

Microsoft Windows [Version 10.0.16299.1508]
(c) 2017 Microsoft Corporation. All rights reserved.

C:\>pip install flask-restful
Collecting flask-restful
  Using cached https://files.pythonhosted.org/packages/17/44/6e490150ee443ca81d5f88b61bb4bbb133d44d75b0b716ebe92489508da
4/Flask_RESTful-0.3.7-py2.py3-none-any.whl
Requirement already satisfied: six>=1.3.0 in c:\python36\lib\site-packages (from flask-restful) (1.12.0)
Requirement already satisfied: pytz in c:\python36\lib\site-packages (from flask-restful) (2016.6)
Requirement already satisfied: Flask>=0.8 in c:\python36\lib\site-packages (from flask-restful) (0.12.2)
Requirement already satisfied: aniso8601>=0.82 in c:\python36\lib\site-packages (from flask-restful) (6.0.0)
Requirement already satisfied: click>=2.0 in c:\python36\lib\site-packages (from Flask>=0.8->flask-restful) (6.6)
Requirement already satisfied: Jinja2>=2.4 in c:\python36\lib\site-packages (from Flask>=0.8->flask-restful) (2.10.1)
Requirement already satisfied: Werkzeug>=0.7 in c:\python36\lib\site-packages (from Flask>=0.8->flask-restful) (0.15.5)
Requirement already satisfied: itsdangerous>=0.21 in c:\python36\lib\site-packages (from Flask>=0.8->flask-restful) (0.2
4)
Requirement already satisfied: MarkupSafe>=0.23 in c:\python36\lib\site-packages (from Jinja2>=2.4->Flask>=0.8->flask-re
stful) (1.1.1)
Installing collected packages: flask-restful
Successfully installed flask-restful-0.3.7
```

Figure 5-7. *Installation of Flask-RESTful*

This command will install the package and all its dependencies.

2. Build the REST API

A RESTful API uses HTTP requests to GET and POST data.

First create a `QuestionAnswering.py` file that will have the question answering code that you have downloaded from GitHub.

3. Deploy Flask REST API

Using Flask, deploy the REST API service and run the following command at the Windows command prompt as shown in Figure 5-8.

```
python QuestionAnswering.py
```

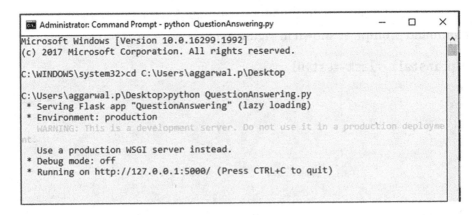

Figure 5-8. *Service deployment*

4. Response from REST API

Now the service has been hosted at the URL http://127.0.0.1:5000/ getResponse. We want the features of the question answering system to be publicly available. Therefore, we will use ngrok to generate a public URL corresponding to the local URL that we configured earlier.

Let's look at the steps to generate a public URL using ngrok.

1. To configure ngrok, download it from https:// ngrok.com/download .

2. The public URL is only available when the auth token is downloaded from https://dashboard. ngrok.com after signing up at https://ngrok.com/ signup.

3. The auth token must be specified to ngrok so that the client is tied to this account. ngrok saves the auth token in ~/.ngrok2/ngrok.yml so that there is no need to repeat the preceding steps.

4. Unzip the downloaded ngrok folder and run the ngrok.exe application.

5. Copy the auth token from the user account mentioned in the command and run this command on the ngrok terminal prompt, as shown in Figure 5-9.

   "ngrok authtoken <AUTHTOKEN>"

c:\Users\   :\Downloads\ngrok-stable-windows-amd64>ngrok authtoken

Figure 5-9. *Token generation*

6. After the previous step, authtoken gets saved to the configuration file, as shown in Figure 5-10.

```
Authtoken saved to configuration file: C:\Users\           /.ngrok2/ngrok.yml
```

***Figure 5-10.** ngrok configuration*

7. ngrok is a command-line application, so type `ngrok http https://<IP>:<PORT>` at this terminal prompt to expose the HTTPS URL. Here the IP and port settings correspond to the question answering API host and port on which the API is hosted, as shown in Figure 5-11.

```
C:\Users\j          \Downloads\ngrok-stable-windows-amd64>ngrok http https://              :
```

***Figure 5-11.** Generate public URL*

8. A new terminal will open after the execution of the command that will show the public URL `https://44e2f215.ngrok.io` corresponding to the local server URL, as shown in Figure 5-12.

```
Web Interface          http://127.0.0.1:4040
Forwarding             http://44e2f215.ngrok.io -> http:/
Forwarding             https://44e2f215.ngrok.io -> http:/

Connections            ttl       opn       rt1       rt5       p50       p90
                       0         0         0.00      0.00      0.00      0.00
```

***Figure 5-12.** Public URL*

Now, you can use the URL highlighted in Figure 5-12. That is, `<URL>/getResponse Flask` is good for a development environment, but not for production. For a production environment, the API should be hosted on

Apache Server. Refer to the following URL to deploy a service on Apache Server in Windows.

https://medium.com/@madumalt/flask-app-deployment-in-windows-apache-server-mod-wsgi-82e1cfeeb2ed

For Linux Server

Prerequisite: Python 3.6.x and Pip need to be installed on your system.

Creation of REST API

1. Install Flask-RESTful

To install, run the following command on Linux Shell as shown in Figure 5-13.

```
$ pip install flask-restful
```

Figure 5-13. *Installation of flask-restful*

This will install the package and its dependencies.

2. Build the REST API

Create an QuestionAnswering.py file that will have the question answering system code that you downloaded from GitHub.

3. Deploy Flask REST API

To deploy the REST API service using Flask, run the following command on Linux Shell, as shown in Figure 5-14.

```
$ python QuestionAnswering.py
```

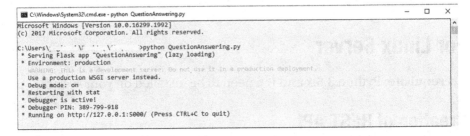

Figure 5-14. *Service deployment*

4. Response from REST API

Now the service has been hosted at the URL `http://127.0.0.1:5000/getResponse`. Because we want features of the question answering system to be publicly available, we use ngrok to generate a public URL corresponding to the local URL that we have configured previously.

Let's look at the steps to generate a public URL using ngrok.

1. To expose a local HTTPS server, download ngrok for Linux server from `https://bin.equinox.io/c/4VmDzA7iaHb/ngrok-stable-linux-amd64.zip`.

2. The public URL is only available when the auth token is downloaded from `https://dashboard.ngrok.com` after signing up at `https://ngrok.com/signup`.

3. The auth token must be specified to ngrok so that the client is bound to this account. ngrok saves the auth token in `~/.ngrok2/ngrok.yml` so that there is no need to repeat this step.

4. To unzip the downloaded ngrok files, run the following command on the terminal as shown in Figure 5-15.

```
$ unzip /path/to/ngrok.zip
```

```
adminuser@airo0002:~$ unzip ngrok-stable-linux-amd64.zip
Archive:  ngrok-stable-linux-amd64.zip
  inflating: ngrok
adminuser@airo0002:~$ ls
```

Figure 5-15. *Unzip ngrok*

5. Copy the auth token from the user account and add in the command. Run this command at the ngrok terminal prompt, as shown in Figure 5-16.

```
"ngrok authtoken <AUTHTOKEN>"
```

```
adminuser@airo0002:~$ ./ngrok authtoken
Authtoken saved to configuration file: /home/adminuser/.ngrok2/ngrok.yml
adminuser@airo0002:~$ 
```

Figure 5-16. *ngrok configuration*

6. After the previous step the auth token will be saved to the configuration file.

7. ngrok is a command-line application, so type ngrok http https://<IP>:<PORT> at this terminal prompt to expose the HTTPS URL. Here the IP and port settings correspond to the question answering API host and port on which the API is hosted, as shown in Figure 5-17.

```
Authtoken saved to configuration file: /home/adminuser/.ngrok2
adminuser@airo0002:~$ ./ngrok http https://            :
```

Figure 5-17. *Generate public URL*

8. After the execution of the command, the
    terminal will be displayed with the public URL
    `https://44e2f215.ngrok.io` corresponding to the
    local server URL as shown in Figure 5-18.

```
Web Interface        http://127.0.0.1:4040
Forwarding           http://44e2f215.ngrok.io -> http://
Forwarding           https://44e2f215.ngrok.io -> http://

Connections          ttl     opn     rt1     rt5     p50     p90
                     0       0       0.00    0.00    0.00    0.00
```

Figure 5-18. *Public URL*

For more details, please refer to the ngrok documentation at `https://ngrok.com/docs`.

Now, you can use the URL as highlighted in Figure 5-18. That is, `<URL>/getResponse` Flask is good for a development environment but not for production. For a production environment, the API should be hosted on Apache Server. Refer to the following URL for guidance on deploying a service on Apache Server in Linux.

`https://www.codementor.io/abhishake/minimal-apache-configuration-for-deploying-a-flask-app-ubuntu-18-04-phu5oa7ft`

Follow the steps given next to release features of a question answering system as a REST API.

1. Create a file named `QuestionAnsweringAPI.py`.

2. Copy the following code and paste it in that file, then save it.

    ```python
    from flask import Flask, request
    import json
    from QuestionAnsweringSystem.QuestionAnswer
    import get_answer_using_bert
    app=Flask(__name__)
    ```

```python
@app.route ("/questionAnswering", methods=['POST'])
def questionAnswering():
    try:
        json_data = request.get_json(force=True)
        query = json_data['query']
        context_list = json_data['context_list']
        result = []
        for val in context_list:
            context = val['context']
            context = context.replace("\n"," ")
            answer_json_final = dict()
            answer = get_answer_using_bert(context, query)
            answer_json_final['answer'] = answer
            answer_json_final['id'] = val['id']
            answer_json_final['question'] = query
            result.append(answer_json_final)
            result={'results':result}
        result = json.dumps(result)
        return result
    except Exception as e:
        return {"Error": str(e)}
if __name__ == "__main__" :
    app.run(port="5000")
```

3. That code processes input passed to an API, calls the function get_answer_using_bert, and sends a response from this function as an API response.

4. Open a command prompt and run the following command.

```
Python QuestionAnsweringAPI.py
```

This will start a service on http://127.0.0.1:5000/ as shown in Figure 5-19.

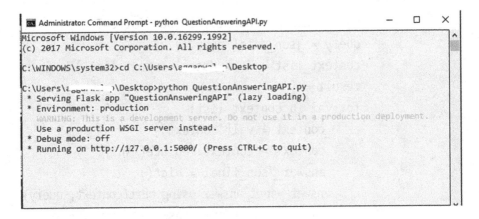

Figure 5-19. *Service deployment.*

5. Now to test the Rest API, we are going to use Postman. This is a REST API client that is used to test the API URL. We can test any complex HTTP/s requests and can also read their responses. First, go to https://www.postman.com/downloads/ to download the Postman tool and install it on your system.

6. After the installation, test following URL and sample request JSON that is being sent to the question answering API end and response JSON that will be received as a response from the API as shown in Figure 5-20.

URL: http://127.0.0.1:5000/questionAnswering

Question answering system sample input request JSON:

```json
{
    "query": "Where was the Football league
    founded?",
    "context_list": [
        {
            "id": 1,
            "context": "In 1888, The Football
            League was founded in England,
            becoming the first of many
            professional football competitions.
            During the 20th century, several of
            the various kinds of football grew
            to become some of the most popular
            team sports in the world"
        }
    ]
}
```

Question answering system sample output response JSON:

```json
{
    "results": [
        {
            "answer": "england",
            "id": 1,
            "question": "Where was the Football
            leagure founded?"
        }
    ]
}
```

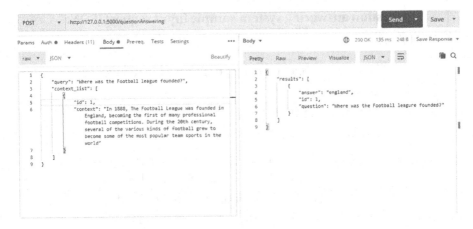

Figure 5-20. *Calling question answering API*

The codebase for this exercise can be downloaded from GitHub at
https://github.com/bertbook/Python_code/tree/master/Chapter5/
QuestionAnsweringSystem.

Open-Domain Question Answering System

An ODQA system aims to find an exact answer to any question from
Wikipedia articles. Thus, for a question, this system will provide a relevant
answer. The default implementation of an ODQA system processes a batch
of queries as an input and returns the answer.

Model Architecture

The architecture of the DeepPavlov ODQA system consists of two
components: a ranker and a reader. To find an answer to any question,
the ranker first retrieves a list of relevant articles from the collection of
documents, and then the reader scans them to identify an answer.

The ranker component is based on the DrQA architecture proposed by
Facebook Research. Specifically, the DrQA approach uses unigram-bigram
hashing and a TF-IDF algorithm to efficiently return a subset of relevant

articles based on a question. The reader component is based on R-NET proposed by Microsoft Research Asia and implemented by Wenxuan Zhou. The R-NET architecture is an end-to-end neural network model that aims to answer questions based on a given document. R-NET first matches the question and the document via gated attention-based recurrent networks to obtain a question-aware document representation. Then, the self-matching attention mechanism refines the representation by matching the document against itself, which effectively encodes information from the whole document. Finally, pointer networks locate the start and end index of the answer in the article. Figure 5-21 shows the logical flow of a DeepPavlov ODQA system.

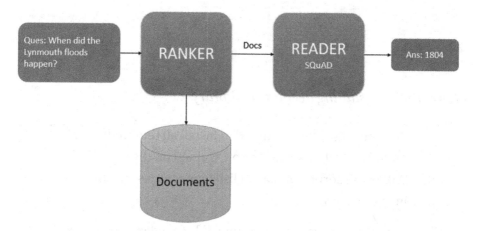

Figure 5-21. The DeepPavlov-based ODQA system architecture

To use this model for an ODQA system, we have used the deeppavlov library in Python. Please note that an ODQA system uses a corpus of Wikipedia articles or documents. Follow these steps to configure and use an ODQA system.

1. Create a new Jupyter notebook and run the
    following command to install the deeppavlov
    library, as shown in Figure 5-22.

    ```
    pip install deeppavlov
    ```

Figure 5-22. *Installing deeppavlov library*

2. Run the following command to install all required
    models, vocabulary, and so on, trained on the
    Wikipedia corpus in the English language, as shown
    in Figure 5-23.

    ```
    ! python -m deeppavlov install en_odqa_infer_wiki
    ```

Note   Please use the '!' symbol before the installation command as
just shown if you are working with Colab Notebook.

Figure 5-23. Installing required packages for deeppavlov

3. Perform the necessary imports required for this implementation as shown here.

```
from deeppavlov import configs
from deeppavlov.core.commands.infer import build_model
```

4. Then we will get an ODQA model using the build_ model class of the deeppavlov library. It takes two arguments:

- **config file path:** Define the name of the config file that contains details of the relevant NLP model to be used. For this case, we will use en_odqa_infer_ wiki. This name implies the ODQA model from Wikipedia.

- **download:** This will be True if the model needs to be downloaded and False otherwise.

```
odqa = build_model(configs.odqa.en_odqa_
infer_wiki, download = True)
```

5. Once the ODQA model has been loaded, you can
    test this model by providing questions such as "Who
    is Virat Kohli?" as shown here.

```
questions = ["Where did guinea pigs
originate?", "Who is virat kohli?"]
answers = odqa(questions)
```

The output of this code will be the answer to questions asked from
Wikipedia documents. Here is the complete code for the ODQA system.

```
from deeppavlov import configs
from deeppavlov.core.commands.infer import build_model

def odqa_deeppavlov(questions):
    odqa = build_model(configs.odqa.en_odqa_infer_wiki,
    download = True)
    results = odqa(questions)
    return results

if __name__ == "__main__" :
        questions = ["Where did guinea pigs originate?", "Who is
        virat kohli?"]
answers = odqa_deeppavlov(questions)
print(answers)
```

Here is the ouput:

```
['Andes of South America',  'Indian international cricketer who
currently captains the India national team']
```

Now, we have seen how an ODQA system can be used for research
or development purposes. Next, consider a scenario where you need
to deploy this feature to be consumed by some website or conversation
system to serve the end user who is looking for an answer to his or her
query. In this case, you need to release or expose features of the ODQA

system as a REST API. Now, follow these steps to release features of the question answering system as a REST API.

1. Create a file named OpenDomainQuestionAnsweringAPI.

2. Copy the following code and paste it in that file, then save it.

```python
from flask import Flask, request
import json
from OpenDomainQuestionAnsweringSystem.OpenDomainQA
import odqa_deeppavlov
app=Flask(__name__)

@route ("/opendomainquestionAnswering",
methods=['POST'])
def opendomainquestionAnswering():
    try:
        json_data = request.get_json(force=True)
        questions = json_data['questions']
        answers_list = odqa_deeppavlov(questions)
        index = 0
        result = []
        for answer in answers_list:
            qa_dict = dict()
            qa_dict['answer']=answer
            qa_dict['question']=questions[index]
            index = index+1
            result.append(qa_dict)
        results = {'results':result}
        results = json.dumps(results)
        return results
```

```
        except Exception as e:
            return {"Error": str(e)}

    if __name__ == "__main__" :
        app.run(debug=True,port="5000")
```

3. This code processes input passed to an API, calls a function odqa_deeppavlov, and sends a response from this function as an API response.

4. Open a command prompt and run the following command.

    Python OpenDomainQuestionAnsweringAPI.py

    This will start a service on http://127.0.0.1:5000/ as shown in Figure 5-24.

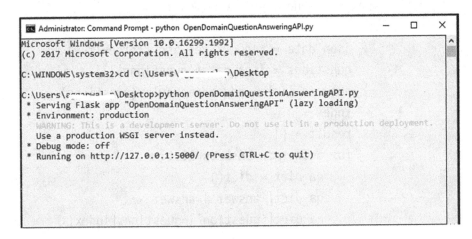

Figure 5-24. *Service deployment*

5. Now, to test this API, Postman can be used. Please
    refer to the following URL and sample request JSON
    that is being provided to the question answering
    API and response JSON that will be received as a
    response from API, as shown in Figure 5-25.

    URL: http://127.0.0.1:5000/opendomain
    questionAnswering

    ODQA system sample input request JSON:

```
{
    "questions": [
        {
            "question": "Where did guinea pigs
            originate?"
        },
{
            "question": "Who is virat kohli?"
        }
    ]
}
```

    ODQA system sample output response JSON:

```
{
    "results": [
        {
            "answer": "Andes of South America",
            "question": "Where did guinea pigs
            originate?"
        },
```

```
{
    "answer": "Indian international
    cricketer who currently captains
    the India national team",
    "question": "Who is virat kohli?"
    }
  ]
}
```

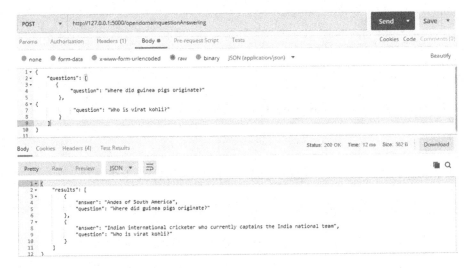

Figure 5-25. *Calling the ODQA system API*

The codebase for this exercise can be downloaded from GitHub at
`https://github.com/bertbook/Python_code/tree/master/Chapter5/`
`OpenDomainQuestionAnsweringSystem`.

DeepPavlov QA System

In the previous section, we discussed how an ODQA system that has
been trained on Wikipedia documents can be used to answer factoid
and non-factoid questions. Next, we look at how DeepPavlov can be

used for an implementation of a contextual-based question answering system where an answer to the question exists in context. As an example, consider the following context and question from a Wikipedia article.

Context: In 1888, The Football League was founded in England, becoming the first of many professional football competitions. During the 20th century, several of the various kinds of football grew to become some of the most popular team sports in the world.

Question: In which year was the Football League founded?

Answer: 1888

Please follow these steps to implement a contextual-based question answering system.

1. Create a new Jupyter notebook and run the following command to install the deeppavlov library.

   ```
   pip install deeppavlov
   ```

2. Run the following command to install all required models, vocabulary, and so on.

   ```
   ! python -m deeppavlov install squad_bert
   ```

Note   Please use the '!' symbol before the installation command as just shown if you are working with Colab Notebook.

---

3. Import the required packages as shown here.

   ```
   from deeppavlov import configs, build_model
   ```

4. Then we will get the BERT model using the `build_
    model` class of the deeppavlov library. It takes two
    arguments:

    - **config file path:** Define the name of the `config` file
      that contains details of the relevant NLP model to
      be used. For this case, we will use `squad_bert`. This
      configuration contains all details for the specific
      BERT model that has been trained over the SQuAD
      dataset.

    - **download:** This is `True` if the model needs to be
      downloaded and `False` otherwise.

      ```
      odqa = build_model(configs.squad.squad_bert,
      download = True)
      ```

5. Once the BERT model has been loaded, you can test
    it by providing a question along with the context to
    extract an answer, as shown here.

    ```
    context = " In 1888, The Football League was
    founded in England, becoming the first of many
    professional football competitions. During the
    20th century, several of the various kinds of
    football grew to become some of the most popular
    team sports in the world."
    ```

    ```
    question = "In which year the Football league was
    founded?"
    ```

    ```
    answers = qa_ deeppavlov (context, question)
    ```

6. The output of this code snippet will be the answer
    extracted from the context for the question asked.

Here is the complete Python code that shows an implementation of a contextual CDQA system.

```python
from deeppavlov import build_model, configs

def qa_deeppavlov(question, context):
    model = build_model(configs.squad.squad_bert,
    download=True)
    result = model([context], [question])
    return result [0]

if __name__=="__main__":

context = "In 1888, The Football League was founded in England,
becoming the first of many professional football competitions.
During the 20th century, several of the various kinds of
football grew to become some of the most popular team sports in
the world."

question = "In which year the Football league was founded?"

answers = qa_deeppavlov (context, question)
            print(answers)
```

Here is the output:

```
1888
```

Now, we have seen how a contextual-based question answering system (another variation of BERT) can be used for research or development purposes. Next, consider a scenario where you need to deploy this feature to be consumed by some website or conversation system to serve the end user who is looking for an answer to his or her query. In this case, you need to release or expose features of the ODQA system as a REST API. Follow the steps given here to release features of the question answering system as a REST API.

1. Create a file named DeepPavlovQASystemAPI.

2. Copy the following code and paste in that file, then save it.

```python
from flask import Flask, request
from DeeppavlovQASystem.QA_Deepplavlov import qa_
deeppavlov
import json
app=Flask(__name__)

@app.route ("/qaDeepPavlov", methods=['POST'])
def qaDeepPavlov():
    try:
        json_data = request.get_json(force=True)
        query = json_data['query']
        context_list = json_data['context_list']
        result = []
        for val in context_list:
            context = val['context']
            context = context.replace("\n"," ")
            answer_json_final = dict()
            answer = qa_deeppavlov(context, query)
            answer_json_final['answer'] = answer
            answer_json_final['id'] = val['id']
            answer_json_final['question'] = query
            result.append(answer_json_final)

        result = json.dumps(result)
        return result
    except Exception as e:
        return {"Error": str(e)}
```

3. This code processes input passed to an API, calls a function qa_deeppavlov, and sends a response from this function as an API response.

4. Open a command prompt and run the following command.

   Python DeepPavlovQASystemAPI.py

   This will start a service on http://127.0.0.1:5000/ as shown in Figure 5-26.

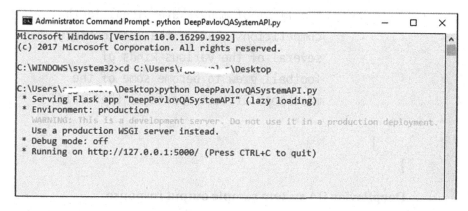

```
Administrator: Command Prompt - python DeepPavlovQASystemAPI.py          —   □   ×
Microsoft Windows [Version 10.0.16299.1992]
(c) 2017 Microsoft Corporation. All rights reserved.

C:\WINDOWS\system32>cd C:\Users\        -\Desktop

C:\Users\              \Desktop>python DeepPavlovQASystemAPI.py
 * Serving Flask app "DeepPavlovQASystemAPI" (lazy loading)
 * Environment: production
   WARNING: This is a development server. Do not use it in a production deployment.
   Use a production WSGI server instead.
 * Debug mode: off
 * Running on http://127.0.0.1:5000/ (Press CTRL+C to quit)
```

Figure 5-26. *Service deployment*

5. Now, to test this API, Postman can be used. Please refer to the following URL and sample request JSON that is being provided to the DeepPavlov QA API end and response JSON that will be received as a response from the API as shown in Figure 5-27.

URL: `http://127.0.0.1:5000/qaDeepPavlov`

DeepPavlov QA system sample input request JSON:

```json
{
    "query": "In which year the Football league
    was founded?",
    "context_list": [
        {
            "id": 1,
            "context": "In 1888, The Football
            League was founded in England, becoming
            the first of many professional football
            competitions. During the 20th century,
            several of the various kinds of
            football grew to become some of the
            most popular team sports in the world"
        }
    ]
}
```

DeepPavlov QA system sample output response JSON:

```json
{
    "results": [
        {
            "answer": "1888",
            "id": 1,
            "question": "In which year the
            Football league was founded?"
        }
    ]
}
```

Figure 5-27. Calling DeepPavlov QA system API

The codebase for this exercise can be downloaded from GitHub at https://github.com/bertbook/Python_code/tree/master/Chapter5/ DeeppavlovQASystem.

Conclusion

This chapter covered the question answering system, which is one of the important applications of the BERT model. We learned about types of question answering Systems like CDQA and ODQA. We built a question answering system using BERT and deployed it as an API for use by a third-party system. In the next chapter, we look at how BERT is used in other NLP tasks.

CHAPTER 6

BERT Model Applications: Other Tasks

In the last chapter, we learned about BERT and its usage in the design of a question answering system. This chapter discusses how BERT can be used for implementation of other NLP tasks such as text classification, named entity recognition, language translation, and more.

BERT has performed well in many benchmark datasets for various NLP tasks such as SQuAD (question answering dataset), Natural Questions (question answering dataset for factoid and non-factoid questions), the IMDB movie review dataset (classification data), and so on. Now, we will see how a BERT-based model trained on these benchmarked datasets can be used as pretrained model for the following NLP tasks.

- Sentiment analysis

- Named entity recognition

- Text classification

- Text summarization

We introduce these topics and then take a looked at their implementation.

© Navin Sabharwal, Amit Agrawal 2021
N. Sabharwal and A. Agrawal, *Hands-on Question Answering Systems with BERT*,
https://doi.org/10.1007/978-1-4842-6664-9_6

Sentiment Analysis

Sentiment analysis is a subfield of NLP that identifies opinions or sentiments of given text across blogs, reviews, news, and so on. It can inform businesses about acceptance of their products and consumer sentiments toward the same. It is also useful to identify hate speech and other issues over social media to identify the population's mood toward a given topic of discussion. Sentiment analysis can even help companies plan product releases on the basis of consumer opinion about particular topics in particular demographic regions.

For this book, we have used a sentiment analysis model that has been trained using BERT, which uses a dataset in .csv format where each data point is pair of sentences and its opinion (i.e., not insult, insult). For inference, the system processes a user's query and provides a sentiment for the same.

Please follow these steps to implement a sentiment analysis system.

1. Create a new Jupyter notebook as covered previously and run the following command to install the deeppavlov library (if you didn't do so in Chapter 5).

    ! pip install deeppavlov

    Once it is installed, you will see the output shown in Figure 6-1.

```
[3] !pip install deeppavlov
```

```
Requirement already satisfied: deeppavlov in /usr/local/lib/python3.6/dist-packages (0.10.0)
Requirement already satisfied: pymorphy2==0.8 in /usr/local/lib/python3.6/dist-packages (from deeppavlov) (0.8)
Requirement already satisfied: pytelegrambotapi==3.6.7 in /usr/local/lib/python3.6/dist-packages (from deeppavlov) (3.6.7)
Requirement already satisfied: pydantic==1.3 in /usr/local/lib/python3.6/dist-packages (from deeppavlov) (1.3)
Requirement already satisfied: ruamel.yaml==0.15.100 in /usr/local/lib/python3.6/dist-packages (from deeppavlov) (0.15.100)
Requirement already satisfied: fastapi==0.47.1 in /usr/local/lib/python3.6/dist-packages (from deeppavlov) (0.47.1)
Requirement already satisfied: scipy==1.4.1 in /usr/local/lib/python3.6/dist-packages (from deeppavlov) (1.4.1)
Requirement already satisfied: rusenttokenize==0.0.5 in /usr/local/lib/python3.6/dist-packages (from deeppavlov) (0.0.5)
Requirement already satisfied: tqdm==4.41.1 in /usr/local/lib/python3.6/dist-packages (from deeppavlov) (4.41.1)
Requirement already satisfied: scikit-learn==0.21.2 in /usr/local/lib/python3.6/dist-packages (from deeppavlov) (0.21.2)
Requirement already satisfied: aio-pika==6.4.1 in /usr/local/lib/python3.6/dist-packages (from deeppavlov) (6.4.1)
Requirement already satisfied: requests==2.22.0 in /usr/local/lib/python3.6/dist-packages (from deeppavlov) (2.22.0)
Requirement already satisfied: sacremoses==0.0.35 in /usr/local/lib/python3.6/dist-packages (from deeppavlov) (0.0.35)
Requirement already satisfied: Cython==0.29.14 in /usr/local/lib/python3.6/dist-packages (from deeppavlov) (0.29.14)
Requirement already satisfied: pymorphy2-dicts-ru in /usr/local/lib/python3.6/dist-packages (from deeppavlov) (2.4.404381.4453942)
Requirement already satisfied: nltk==3.4.5 in /usr/local/lib/python3.6/dist-packages (from deeppavlov) (3.4.5)
Requirement already satisfied: pyopenssl==19.1.0 in /usr/local/lib/python3.6/dist-packages (from deeppavlov) (19.1.0)
Requirement already satisfied: pytz==2019.1 in /usr/local/lib/python3.6/dist-packages (from deeppavlov) (2019.1)
Requirement already satisfied: overrides==2.7.0 in /usr/local/lib/python3.6/dist-packages (from deeppavlov) (2.7.0)
Requirement already satisfied: uvicorn==0.11.1 in /usr/local/lib/python3.6/dist-packages (from deeppavlov) (0.11.1)
Requirement already satisfied: pandas==0.25.3 in /usr/local/lib/python3.6/dist-packages (from deeppavlov) (0.25.3)
Requirement already satisfied: h5py==2.10.0 in /usr/local/lib/python3.6/dist-packages (from deeppavlov) (2.10.0)
Requirement already satisfied: docopt==0.6.6 in /usr/local/lib/python3.6/dist-packages (from pymorphy2==0.8->deeppavlov) (0.6.2)
Requirement already satisfied: dawg-python>=0.7 in /usr/local/lib/python3.6/dist-packages (from pymorphy2==0.8->deeppavlov) (0.7.2)
Requirement already satisfied: pymorphy2-dicts<3.0,>=2.4 in /usr/local/lib/python3.6/dist-packages (from pymorphy2==0.8->deeppavlov) (2.4.393442.3710985)
Requirement already satisfied: six in /usr/local/lib/python3.6/dist-packages (from pytelegrambotapi==3.6.7->deeppavlov) (1.12.0)
Requirement already satisfied: dataclasses>=0.8; python_version < "3.7" in /usr/local/lib/python3.6/dist-packages (from pydantic==1.3->deeppavlov) (0.7)
Requirement already satisfied: starlette<0.12.9,>=0.12.9 in /usr/local/lib/python3.6/dist-packages (from fastapi==0.47.1->deeppavlov) (0.12.9)
Requirement already satisfied: joblib>=0.11 in /usr/local/lib/python3.6/dist-packages (from scikit-learn==0.21.2->deeppavlov) (0.15.1)
Requirement already satisfied: aiormq<4,>=3.2.0 in /usr/local/lib/python3.6/dist-packages (from aio-pika==6.4.1->deeppavlov) (3.2.3)
```

Figure 6-1. *Installation of deeppavlov*

2. Because we are going to use sentiment analysis
    we we will use a model that has been trained on
    sentiment data. Run the following command to
    download a trained model, `insults_kaggle_bert`.

    `! python -m deeppavlov install insults_kaggle_bert`

Note   Please use the '!' symbol before the installation command as
just shown if you are working with Colab Notebook.

Once it is installed, you will see the output shown in Figure 6-2.

Figure 6-2. *Installation of packages*

3. Perform the necessary imports as required for this
    implementation using this command.

```
from deeppavlov import build_model, configs
```

4. Then we will get a sentiment analysis model using
    the build_model class of the deeppavlov library. It
    takes two arguments:

    •   **config file path:** Define the name of the config file
        that contains details of the relevant NLP model to
        be used. For this case, we will use insults_kaggle_
        bert. This contains the configuration required to
        use the sentiment model.

    •   **download:** This is True if the model needs to be
        downloaded, and False otherwise. Because we
        are doing this for the first time, the value of this
        argument will be True.

```
sentiment_model = build_model(configs.classifiers.
insults_kaggle_bert, download=True)
```

5. Once the sentiment model has been loaded, you can test this model by asking questions such as "You are so dumb!," "This movie is good," and so on, and passing these questions as an argument to the sentiment_model function shown here.

```
test_input = ['This movie is good', 'You are so dumb!']
results = sentiment_model(test_ input)
```

The output of this code segment will be Not Insult or Insult depending on the question asked. Here is the complete end-to-end codebase to use sentiment analysis.

```
from deeppavlov import build_model, configs

def build_sentiment_model ():
    model = build_model(configs.classifiers.insults_kaggle_
    bert, download=True)
    return model

test_input = ['This movie is good', 'You are so dumb!']

if __name__ == "__main__" :
        sentiment_model = build_sentiment_model()
        results = sentiment_model(test_ input)
        print(results)
```

This is the output:

```
['Not Insult', 'Insult']
```

Now that we have seen how a sentiment analysis system based on BERT can be leveraged for research purposes, let's consider a scenario where you need to enable sentiment analysis in a conversation system such that it can identify sentiments of a user based on a query or a response given by user. This would help the conversation system to respond on the basis of the sentiments of a user. Follow the steps given here to release features of sentiment analysis system as a REST API.

1. Create a file named SentimentAnalysisAPI.py.

2. Copy the code shown here and paste it in this file, then save it.

```
from flask import Flask, request
import json
from SentimentAnalysis.SentimentAnalysis
import build_sentiment_model
app=Flask(__name__)

@app.route ("/sentimentAnalysis",
methods=['POST'])
def sentimentAnalysis():
    try:
        json_data = request.get_json(force=True)
        questions = json_data['questions']
        sentiment_model = build_sentiment_
        model()
        questions_list =[]
        for ques in questions:
            questions_list.append(ques)

        model_output = sentiment_model
        (questions_list)
        index = 0
        result = []
        for ans in model_output:
            sentiment_qa =dict()
            sentiment_qa['qustion'] =
            questions_list[index]
```

```
            sentiment_qa['answer'] = ans
            result.append(sentiment_qa)

        result={'results':result}
        result = json.dumps(result)
        return result
    except Exception as e:
        return {"Error": str(e)}

if __name__ == "__main__" :
    app.run(port="5000")
```

3. This code processes input passed to an API, calls the build_sentiment_model function, and sends a response from this function as an API response.

4. Open a command prompt and run the following command.

```
Python SentimentAnalysisAPI.py
```

This will start a service on http://127.0.0.1:5000/ as shown in Figure 6-3.

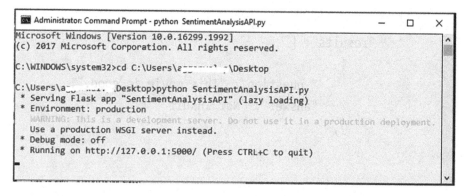

Figure 6-3. *Service deployment*

5. Now, to test the Rest API, Postman can be used.
    Please refer to the URL and sample request JSON
    that is being provided to the sentiment analysis
    API and response JSON that will be received as a
    response from the API as shown in Figure 6-4.

URL: `http://127.0.0.1:5000/sentimentAnalysis`

Sentiment analysis system sample input request JSON:

```
{
    "questions": [
        {
            "question": "This movie is good."
        },
        {
            "question": "You are so dumb!"
        }
    ]
}
```

Sentiment analysis system sample output response JSON:

```
{
    "results": [
        {
            " question": "This movie is good.",
            "answer": "Not Insult"
        },
```

```
{
    " question": "You are so dumb!",
    "answer": "Insult"
}
]
}
```

Figure 6-4. *Calling sentiment analysis system API*

The codebase for this exercise can be downloaded from GitHub at
`https://github.com/bertbook/Python_code/tree/master/Chapter6/`
`SentimentAnalysis`.

Named Entity Recognition

Named entity recognition is a subfield of information extraction where it
aims to extract nouns or noun phrases from text data and classify them
into categories as such as person, place, time, organization, and so on.

This is mainly useful for conversion of unstructured text to structured text. Entity recognition plays a major role in the following systems.

- **Search engine**: This is used to identify relevant documents for a query asked by the user. For an example, let's use "What is Microsoft Outlook?" In this query, "Microsoft Outlook" is an entity of type application. The search engine will thus give more importance to a document where Microsoft Outlook is being identified as an entity.

- **Conversation system**: Entity plays a major role in the design of a conversation system. Entities are being used in conversation systems to disambiguate a question asked by the user if it is related to common issues but for different entities. As an example, a user has entered the query "I am facing an issue in Outlook." The conversation system has two solutions: one for Outlook and the other for Gmail. Because Outlook and Gmail both are different entities, so are the solutions. Therefore, after identification of intent (i.e., Issue), the next identification will be entity (i.e., Outlook) and the conversation system provides a solution accordingly.

There exist many annotated datasets for entity recognition. For this book, though, we will demonstrate an entity model that has been trained on the OntoNotes dataset using BERT as a baseline. This dataset is a collection of 1,745,000 English, 900,000 Chinese, and 300,000 Arabic text data collected from a range of sources such as telephone conversations, newswire, broadcast news, broadcast conversation, and blogs.

In this dataset, entities have been annotated with 18 categories such as organization, art work, numbers in word, numbers, quantity, person, location, geopolitical entity, time, date, facility, event, law, nationalities or religious or political groups, language, currency, percentage, and product, among others.

In this section, we explore how a named entity recognition system that has been trained on an OntoNotes dataset using BERT can be used. Please follow these steps to implement a named entity recognition system.

1. Create a new Jupyter notebook, as mentioned previously, and run the following command to install the deeppavlov library.

   ```
   ! pip install deeppavlov
   ```

   Once installed, you will see output that looks like Figure 6-5.

```
[3] !pip install deeppavlov
   Requirement already satisfied: deeppavlov in /usr/local/lib/python3.6/dist-packages (0.10.0)
   Requirement already satisfied: pymorphy2==0.8 in /usr/local/lib/python3.6/dist-packages (from deeppavlov) (0.8)
   Requirement already satisfied: pytelegrambotapi==3.6.7 in /usr/local/lib/python3.6/dist-packages (from deeppavlov) (3.6.7)
   Requirement already satisfied: pydantic==1.3 in /usr/local/lib/python3.6/dist-packages (from deeppavlov) (1.3)
   Requirement already satisfied: ruamel.yaml==0.15.100 in /usr/local/lib/python3.6/dist-packages (from deeppavlov) (0.15.100)
   Requirement already satisfied: fastapi==0.47.1 in /usr/local/lib/python3.6/dist-packages (from deeppavlov) (0.47.1)
   Requirement already satisfied: scipy==1.4.1 in /usr/local/lib/python3.6/dist-packages (from deeppavlov) (1.4.1)
   Requirement already satisfied: rusenttokenize==0.0.5 in /usr/local/lib/python3.6/dist-packages (from deeppavlov) (0.0.5)
   Requirement already satisfied: tqdm==4.41.1 in /usr/local/lib/python3.6/dist-packages (from deeppavlov) (4.41.1)
   Requirement already satisfied: scikit-learn==0.21.2 in /usr/local/lib/python3.6/dist-packages (from deeppavlov) (0.21.2)
   Requirement already satisfied: aio-pika==6.4.1 in /usr/local/lib/python3.6/dist-packages (from deeppavlov) (6.4.1)
   Requirement already satisfied: requests==2.22.0 in /usr/local/lib/python3.6/dist-packages (from deeppavlov) (2.22.0)
   Requirement already satisfied: sacremoses==0.0.35 in /usr/local/lib/python3.6/dist-packages (from deeppavlov) (0.0.35)
   Requirement already satisfied: numpy==1.18.0 in /usr/local/lib/python3.6/dist-packages (from deeppavlov) (1.18.0)
   Requirement already satisfied: Cython==0.29.14 in /usr/local/lib/python3.6/dist-packages (from deeppavlov) (0.29.14)
   Requirement already satisfied: nltk==3.4.5 in /usr/local/lib/python3.6/dist-packages (from deeppavlov) (3.4.5)
   Requirement already satisfied: pyopenssl==19.1.0 in /usr/local/lib/python3.6/dist-packages (from deeppavlov) (19.1.0)
   Requirement already satisfied: pytz==2019.1 in /usr/local/lib/python3.6/dist-packages (from deeppavlov) (2019.1)
   Requirement already satisfied: overrides==2.7.0 in /usr/local/lib/python3.6/dist-packages (from deeppavlov) (2.7.0)
   Requirement already satisfied: uvicorn==0.11.1 in /usr/local/lib/python3.6/dist-packages (from deeppavlov) (0.11.1)
   Requirement already satisfied: pandas==0.25.3 in /usr/local/lib/python3.6/dist-packages (from deeppavlov) (0.25.3)
   Requirement already satisfied: h5py==2.10.0 in /usr/local/lib/python3.6/dist-packages (from deeppavlov) (2.10.0)
   Requirement already satisfied: docopt>=0.6 in /usr/local/lib/python3.6/dist-packages (from pymorphy2==0.8->deeppavlov) (0.6.2)
   Requirement already satisfied: dawg-python>=0.7 in /usr/local/lib/python3.6/dist-packages (from pymorphy2==0.8->deeppavlov) (0.7.2)
   Requirement already satisfied: pymorphy2-dicts<3.0,>=2.4 in /usr/local/lib/python3.6/dist-packages (from pymorphy2==0.8->deeppavlov) (2.4.393442.3710985)
   Requirement already satisfied: six in /usr/local/lib/python3.6/dist-packages (from pytelegrambotapi==3.6.7->deeppavlov) (1.12.0)
   Requirement already satisfied: dataclasses>=0.6; python_version < "3.7" in /usr/local/lib/python3.6/dist-packages (from pydantic==1.3->deeppavlov) (0.7)
   Requirement already satisfied: starlette<0.12.9,>=0.12.9 in /usr/local/lib/python3.6/dist-packages (from fastapi==0.47.1->deeppavlov) (0.12.9)
   Requirement already satisfied: joblib>=0.11 in /usr/local/lib/python3.6/dist-packages (from scikit-learn==0.21.2->deeppavlov) (0.15.1)
   Requirement already satisfied: aiormq~=3.2.0 in /usr/local/lib/python3.6/dist-packages (from aio-pika==6.4.1->deeppavlov) (3.2.2)
```

Figure 6-5. *Installation of deeppavlov*

2. We are going to use an entity recognition system that has been trained on OntoNotes data as shown in Figure 6-6. Hence, run the following command to download the trained model, ner_ontonotes_bert_mult.

   ```
   ! python -m deeppavlov install ner_ontonotes_
   bert_mult
   ```

Note   Please use the '!' symbol before the installation command as just shown if you are working with Colab Notebook.

***Figure 6-6.** Installation of packages*

3. Perform the necessary imports as required for this implementation using this command.

```
from deeppavlov import build_model, configs
```

4. We will then get an entity model using the build_ model class of the deeppavlov library. It takes two arguments:

   - **config file path:** Define the name of the config file that contains the details of the relevant NLP model to be used. For this case, we will use ner_ontonotes_bert_mult. This file contains all configurations required for the entity model trained on OntoNotes.

- **download:** This is True if the model needs to be downloaded and False otherwise. Because we are doing this for the first time, the value of this argument will be True.

```
ner_model = build_model(configs. ner.ner_
ontonotes_bert_mult, download=True)
```

5. Once the entity recognition model has been loaded, you can test this model by providing text such as "Amazon rainforests are located in South America." and passing it as an argument to the function named ner_model as shown here.

```
test_input = ["Amazon rainforests are located
in South America."]
results = ner_model(test_ input)
```

The output of these code snippets contains words along with their tagged entities, as shown in Figure 6-7.

	Word	Entity
0	Amazon	B-LOC
1	rainforests	O
2	are	O
3	located	O
4	in	O
5	South	B-LOC
6	America	I-LOC
7	.	O

Figure 6-7. *Named entity recognition system result*

Here is the complete Python code for this implementation.

```
from deeppavlov import build_model, configs
import pandas as pd
def build_ner_model ():
    model = build_model(configs. ner.ner_ontonotes_bert_mult,
    download=True)
    return model

if __name__=="__main__":

  test_input = ["Amazon rainforests are located in South
  America."]
  ner_model = build_ner_model()
  results = ner_model(test_ input)
  results = pd.DataFrame(zip(results[0][0],results[1][0]),
  columns=['Word','Entity'])
  print(results)
```

The output is the recognized entities as shown in Figure 6-8.

	Word	Entity
0	Amazon	B-LOC
1	rainforests	O
2	are	O
3	located	O
4	in	O
5	South	B-LOC
6	America	I-LOC
7	.	O

Figure 6-8. *Named entity recognition system result*

Now that we have seen how an entity recognition system based on BERT can be used for research purposes, we next consider a scenario where we need to deploy this feature to be consumed by a conversation system. A conversation system generally uses entities to configure or develop use cases. As an example, for use case "Facing an issue with Outlook," this system can be used to identify "Outlook" as an entity. In this case, you need to release or expose the features of the entity recognition system as a REST API using the following steps.

1.   Create a file with named NamedEntityAPI.

2.   Copy the following code and paste in that file, then save it.

```
from flask import Flask, request
import json
from NamedEntityRecognition.
NamedEntityRecognition import build_ner_model
app=Flask(__name__)

@app.route ("/namedEntity",
methods=['POST'])
def namedEntity():
    try:
        json_data = request.get_json(force=True)
        query = json_data['query']
        ner_model = build_ner_model()
        model_output = ner_model([query])
        words= model_output[0][0]
        tags = model_output[1][0]
        result_json = dict()
        result_json['query'] = query
```

```
            entities = []
            index = 0

            for word in words:
                word_tag_dict=dict()
                word_tag_dict['word'] = word
                word_tag_dict['tag'] = tags[index]
                index = index+1
                entities.append(word_tag_dict)

            result_json['entities'] = entities
            result = json.dumps(result_json)
            return result
        except Exception as e:
            return {"Error": str(e)}

if __name__ == "__main__" :
    app.run(port="5000")
```

3. This code processes the input passed to an API, calls the build_ner_model function, and sends a response from this function as an API response.

4. Open a command prompt and run the following command.

    `Python NamedEntiityAPI.py`

    This will start a service on `http://127.0.0.1:5000/` as shown in Figure 6-9.

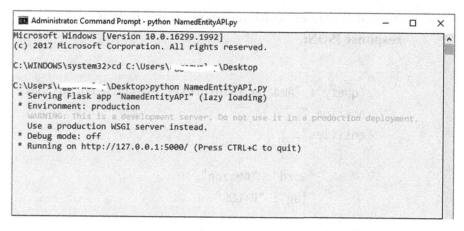

```
Administrator: Command Prompt - python NamedEntityAPI.py                    —    □    ×

Microsoft Windows [Version 10.0.16299.1992]
(c) 2017 Microsoft Corporation. All rights reserved.

C:\WINDOWS\system32>cd C:\Users_____ _\Desktop

C:\Users_____ _\Desktop>python NamedEntityAPI.py
 * Serving Flask app "NamedEntityAPI" (lazy loading)
 * Environment: production
   WARNING: This is a development server. Do not use it in a production deployment.
   Use a production WSGI server instead.
 * Debug mode: off
 * Running on http://127.0.0.1:5000/ (Press CTRL+C to quit)
```

Figure 6-9. *Service deployment*

5. Now, to test this API, Postman can be used as
    explained in Chapter 5. Please refer to the following
    URL and sample request JSON that is being
    provided to the named entity recogntion system
    API and response JSON that will be received as a
    response from the API as shown in Figure 6-10.

    URL: http://127.0.0.1:5000/namedEntity
    **Named entity recognition system sample input
    request JSON:**

    ```
    {
        "query": "Amazon rainforests are located in South
        America."
    }
    ```

**Named entity recognition system sample output
response JSON:**

```json
{
    "query": "Amazon rainforests are located in South
    America.",
    "entities": [
        {
            "word": "Amazon",
            "tag": "B-LOC"
        },
        {
            "word": "South",
            "tag": "B-LOC "
        },
    {

            "word": "America",
            "tag": "I-LOC "
        }
    ]
}
```

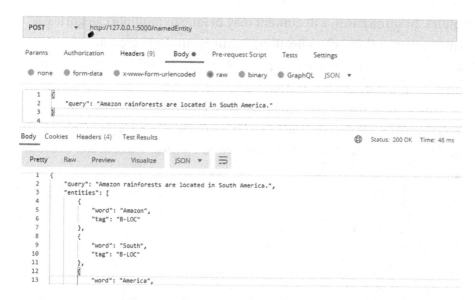

Figure 6-10. *Calling named entity recognition system API*

The codebase for this exercise can be downloaded from GitHub at
`https://github.com/bertbook/Python_code/tree/master/Chapter6/`
`NamedEntityRecognition`.

Text Classification

Text classification can be defined as the problem of assigning or
categorizing text into a particular category or class. Document
classification or categorization, intent classification, spam blog detection,
and more falls under text classification. Here, text can be anything such
as a sentence, a document, blogs, and so on. Text classification leverages
NLP methods for preprocessing such as tokenization, stop-word removal,
phrase extraction, entity extraction, and so on.

During inference, text classification analyzes the text (document,
blog, or sentence) and assigns it to pretrained categories. As an example,
if document is referring to politics, then this belongs to the category of

politics. In some cases, a document may belong to multiple categories (known as multilabel classification). As an example, if document is talking about politics as well as sports, then it will be classified into both categories; that is, politics and sports.

This section shows how a text categorization system trained on newsgroup datasets using BERT can be used. Here, we are going to classify news articles into their respective categories. This dataset has four categories for news articles:

- alt.atheism

- soc.religion.christian

- comp.graphics

- sci.med

We will use ktrain and tensorflow_gpu for this implementation. Please note that this implementation requires the GPU version of TensorFlow to be installed on the system. Therefore, please ensure you have a GPU-enabled system.

1. Create a new Jupyter notebook as mentioned previously and run the following command to install tensorflow_gpu and the ktrain library.

```
! pip3 install -q tensorflow_gpu==2.1.0
!pip3 install -q ktrain
```

After successful installation of the package, it shows an output as displayed in Figure 6-11.

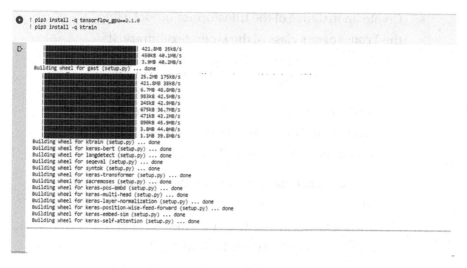

```
! pip3 install -q tensorflow_gpu==2.1.0
! pip3 install -q ktrain
```

```
                                    421.8MB 35kB/s
                                    450kB 40.1MB/s
                                    3.9MB 40.2MB/s
Building wheel for gast (setup.py) ... done
                                    25.2MB 175kB/s
                                    421.8MB 38kB/s
                                    6.7MB 48.8MB/s
                                    983kB 42.5MB/s
                                    245kB 42.9MB/s
                                    675kB 36.7MB/s
                                    471kB 43.2MB/s
                                    896kB 45.9MB/s
                                    3.8MB 44.0MB/s
                                    1.1MB 39.8MB/s
Building wheel for ktrain (setup.py) ... done
Building wheel for keras-bert (setup.py) ... done
Building wheel for langdetect (setup.py) ... done
Building wheel for seqeval (setup.py) ... done
Building wheel for syntok (setup.py) ... done
Building wheel for keras-transformer (setup.py) ... done
Building wheel for sacremoses (setup.py) ... done
Building wheel for keras-pos-embd (setup.py) ... done
Building wheel for keras-multi-head (setup.py) ... done
Building wheel for keras-layer-normalization (setup.py) ... done
Building wheel for keras-position-wise-feed-forward (setup.py) ... done
Building wheel for keras-embed-sim (setup.py) ... done
Building wheel for keras-self-attention (setup.py) ... done
```

Figure 6-11. *Installation of TensorFlow*

2. Import the packages required for this implementation,
    such as the fetch_20newsgroup dataset from sklearn
    and the ktrain library as shown here.

    ```
    from sklearn.datasets import fetch_20newsgroups
    import ktrain
    ```

3. Next download and retrieve the fetch_20newsgroup
    dataset for only four categories: alt.atheism, soc.
    religion.christian, comp.graphics, and sci.med.
    Divide them into a training and test set with
    shuffling enabled, as shown here.

    ```
    classes = ['alt.atheism', 'soc.religion.
    christian','comp.graphics', 'sci.med']
    train_data = fetch_20newsgroups(subset='train',
    categories=classes, shuffle=True, random_state=42)
    test_data = fetch_20newsgroups(subset='test',
    categories=classes, shuffle=True, random_state=42)
    ```

4. Create an instance of the transformer model using the Transformer class of the ktrain.text library. It requires values of some of the parameters to be defined as shown here.

- **Model name:** This indicates the name of the BERT model to be used. In this case, we have used distillBERT instead of BERT base.

- **Length of article:** This sets the maximum length of an article. Here, maximum length can only be 512. If you specific any article of a length greater than 512, it will automatically be truncated.

- **Classes:** This is a list of classes that needs to be considered for training.

5. The next step is to preprocess training and test data to generate their embedded representation of articles using distillBERT. Pass these data and the model to the get_learner function of ktrain to get an instance of the classification model with all configuration parameters, such as batch_size, instance of a model, training data, and test data.

```
MODEL_NAME = 'distilbert-base-uncased'
trans = text.Transformer(MODEL_NAME,
maxlen=500, classes=train_classes)
train_preprocess = trans.preprocess_
train(train_features, train_labels)
val_preprocess = trans.preprocess_test(test_
features, test_labels)
model_data = trans.get_classifier()
```

```
classification_model = ktrain.get_
learner(model_data, train_data=train_
preprocess, val_data=val_preprocess, batch_
size=6)
classification_model.fit_onecycle(5e-5, 4)
```

6. Once the classification model has been trained, then
    this model can be tested on unseen data, as shown
    here.

```
predictor = ktrain.get_
predictor(classification_model.model,
preproc=trans)
input_text = 'Babies with down syndrome have an
extra chromosome.'
results = predictor.predict(input_text)
```

Here is the complete Python code to implement text classification.

```
from sklearn.datasets import fetch_20newsgroups
import ktrain
from ktrain import text

def preprocess_dataset():
    classes = ['alt.atheism', 'soc.religion.christian',
    'comp.graphics', 'sci.med']
    train_data = fetch_20newsgroups(subset='train',
    categories=classess, shuffle=True, random_state=42)
    test_data = fetch_20newsgroups(subset='test',
    categories=classes, shuffle=True, random_state=42)
    return train_data.data,train_data.target, test_data.
    data,test_data.target,classes
```

```python
def create_text_classification_model():
    MODEL_NAME = 'distilbert-base-uncased'
    train_features, train_labels, test_features, test_labels,
    train_classes =preprocess_dataset()
    trans = text.Transformer(MODEL_NAME, maxlen=500,
    classes=train_classes)
    train_preprocess = trans.preprocess_train(train_features,
    train_labels)
    val_preprocess = trans.preprocess_test(test_features,
    test_labels)
    model_data = trans.get_classifier()
    classification_model = ktrain.get_learner(model_data,
    train_data=train_preprocess, val_data=val_preprocess,
    batch_size=6)
    classification_model.fit_onecycle(5e-5, 4)
    return classification_model, trans

def predict_category(classification_model, trans, input_text):
    predictor = ktrain.get_predictor(classification_model.
    model, preproc=trans)
    results = predictor.predict(input_text)
    return results

if __name__ == "__main__" :

        classification_model, trans = create_text_
        classification_model()
        input_text = 'Babies with down syndrome have an extra
        chromosome.'
        print(predict_category(classification_model, trans,
        input_text))
```

As you can see from the following output, for a text "Babies with down syndrome have an extra chromosome." The category is sci.med.

```
sci.med
```

Now, we have seen how a text classification system based on BERT can be used for research purposes. Next, consider a scenario where you need to deploy this feature to be consumed by a conversation system. A conversation system can leverage this as intent classification or a recognition system to configure or develop use cases. As an example, for a use case "Facing an issue with Outlook," this system can be used to identify an intent as "Issue." In this case, you need to release or expose features of the intent classification system as a REST API by following these steps.

1. Create a file named `TextClassificationAPI`.

2. Copy the following code and paste it in that file, then save it.

```
from flask import Flask, request
import json
from TextClassification.TextClassification
import create_text_classification_model,
predict_category
from TextClassification import create_text_
classification_model
app=Flask(__name__)
result={}

@app.route ("/textClassification",
methods=['POST'])
def textClassification ():
    try:
        json_data = request.get_json(force=True)
```

```
        input_text = json_data['query']

        classification_model, trans =
        create_text_classification_model()
        category = predict_category
        (classification_model, trans,
        input_text)

        result = {}
        result['query'] = input_text
        result['category'] = category

        result = json.dumps(result)
        return result

    except Exception as e:
        error = {"Error": str(e)}
        error = json.dumps(error)
        return error

if __name__ == "__main__" :
    app.run(port="5000")
```

3. This code processes the input passed to an API, calls
    the create_text_classification_model function,
    and sends a response from this function as an API
    response.

4. Open a command prompt and run the following
    command.

    `Python TextClassificationAPI.py`

    This will start a service on http://127.0.0.1:5000/ as
    shown in Figure 6-12.

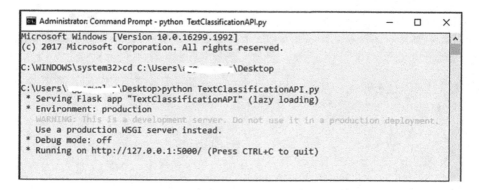

Figure 6-12. Service deployment

5. Now, to test the Rest API, Postman can be used, as
    mentioned in Chapter 5. Please refer the following
    URL and sample request JSON that is being
    provided to the text classification API and response
    JSON that will be received as a response from the
    API, as shown in Figure 6-13.

URL: http://127.0.0.1:5000/textClassification
Text classification system sample input request JSON:

```
{

        "query": "Babies with down syndrome have an extra
        chromosome."

}
```

Text classification system sample output response JSON:

```
{

    "query": "Babies with down syndrome have an extra
    chromosome.",
    "category": "sci.med"

}
```

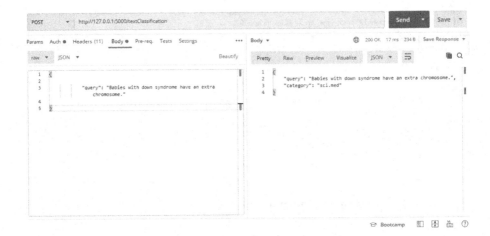

Figure 6-13. *Calling text classification system API*

The codebase for this exercise can be downloaded from GitHub at
`https://github.com/bertbook/Python_code/tree/master/Chapter6/`
`TextClassification`.

Text Summarization

Text summarization is a process that uses NLP and NLU to generate or
extract a summary out of a document while preserving the actual meaning
of the document. In other words, the summary should be very similar to
what the document says. This function has been quite popular in search
engine systems, where a document presented to a user also includes a
summary of the document instead of the entire document text. Document
summarization can be single or multidocument summarization. Text
summarization problems can be classified into two types:

- **Extractive summarization**: In extractive
  summarization, sentences in the generated summary
  will be only from the document itself. There won't be
  any modification to sentences in the summary. This

can also be defined as rearrangement of sentences on the basis of their relevance to document topics. Several approaches such as TF-IDF, cosine similarity, graph-based approaches, entity extraction, tokenization, and so on have been used to actively develop document summarization systems.

- **Abstractive summarization:** In abstractive summarization, sentences in the generated summary won't be actual sentences from the document itself. These sentences will be modified based on language semantics used in the document. Various neural network–based approaches such as LSTM, GRU, and so on have been used to implement this.

In this section, we discuss how BERT is being used to generate a summary of a document. BERT proposes a new architecture known as BERTSUM that generates a summary from a document. As usual, BERT is used to generate embedding of multiple sentences where the token [CLS] is inserted before the start of the first sentence followed by other sentences that have been separated by the token [SEP]. Next, segment and positional embedding have been appended to segregate between sentences. Then these sentence vectors pass through the summarization layer to select representative sentences for the summary. In the summarization layer, any neural network can construct the summary. Figure 6-14 shows the document summarization model architecture.

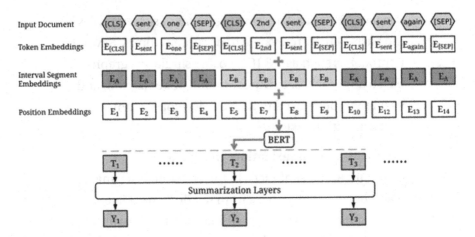

Figure 6-14. *Architecture of a BERTSUM model*

Now, let's look at how a BERT-based extractive document summarization model can be leveraged. We use `bert-extractive-summarizer`, one of the implementations of extractive document summarization in Python to demonstrate the same.

1. Create a new Jupyter notebook as mentioned previously and run the following command to install `bert-extractive-summarizer`.

   ```
   ! pip3 install bert-extractive-summarizer
   ```

   After successful installation of the package, the output shown in Figure 6-15 is displayed.

```
[8] ! pip3 install bert-extractive-summarizer

 Collecting bert-extractive-summarizer
    Downloading https://files.pythonhosted.org/packages/23/1d/71f0a5c7f81b1a87d442da6a93Se9ddeb5e462e41512962e11bd10533cd9/bert-extractive-summarizer-0.4.2.tar.gz
  Requirement already satisfied: transformers in /usr/local/lib/python3.6/dist-packages (from bert-extractive-summarizer) (2.11.0)
  Requirement already satisfied: scikit-learn in /usr/local/lib/python3.6/dist-packages (from bert-extractive-summarizer) (0.21.3)
  Requirement already satisfied: spacy in /usr/local/lib/python3.6/dist-packages (from bert-extractive-summarizer) (2.2.4)
  Requirement already satisfied: sacremoses in /usr/local/lib/python3.6/dist-packages (from transformers->bert-extractive-summarizer) (0.0.43)
  Requirement already satisfied: tokenizers==0.7.0 in /usr/local/lib/python3.6/dist-packages (from transformers->bert-extractive-summarizer) (0.7.0)
  Requirement already satisfied: filelock in /usr/local/lib/python3.6/dist-packages (from transformers->bert-extractive-summarizer) (3.0.12)
  Requirement already satisfied: regex!=2019.12.17 in /usr/local/lib/python3.6/dist-packages (from transformers->bert-extractive-summarizer) (2019.12.20)
  Requirement already satisfied: requests in /usr/local/lib/python3.6/dist-packages (from transformers->bert-extractive-summarizer) (2.23.0)
  Requirement already satisfied: packaging in /usr/local/lib/python3.6/dist-packages (from transformers->bert-extractive-summarizer) (20.4)
  Requirement already satisfied: dataclasses; python_version < "3.7" in /usr/local/lib/python3.6/dist-packages (from transformers->bert-extractive-summarizer) (0.7)
  Requirement already satisfied: numpy in /usr/local/lib/python3.6/dist-packages (from transformers->bert-extractive-summarizer) (1.18.5)
  Requirement already satisfied: tqdm>=4.27 in /usr/local/lib/python3.6/dist-packages (from transformers->bert-extractive-summarizer) (4.41.1)
  Requirement already satisfied: sentencepiece in /usr/local/lib/python3.6/dist-packages (from transformers->bert-extractive-summarizer) (0.1.91)
  Requirement already satisfied: joblib>=0.11 in /usr/local/lib/python3.6/dist-packages (from scikit-learn->bert-extractive-summarizer) (0.15.1)
  Requirement already satisfied: scipy>=0.17.0 in /usr/local/lib/python3.6/dist-packages (from scikit-learn->bert-extractive-summarizer) (1.4.1)
  Requirement already satisfied: wasabi<1.1.0,>=0.4.0 in /usr/local/lib/python3.6/dist-packages (from spacy->bert-extractive-summarizer) (0.7.0)
  Requirement already satisfied: plac<1.2.0,>=0.9.6 in /usr/local/lib/python3.6/dist-packages (from spacy->bert-extractive-summarizer) (1.1.3)
  Requirement already satisfied: preshed<3.1.0,>=3.0.2 in /usr/local/lib/python3.6/dist-packages (from spacy->bert-extractive-summarizer) (3.0.2)
  Requirement already satisfied: thinc==7.4.0 in /usr/local/lib/python3.6/dist-packages (from spacy->bert-extractive-summarizer) (7.4.0)
  Requirement already satisfied: setuptools in /usr/local/lib/python3.6/dist-packages (from spacy->bert-extractive-summarizer) (47.3.1)
  Requirement already satisfied: murmurhash<1.1.0,>=0.28.0 in /usr/local/lib/python3.6/dist-packages (from spacy->bert-extractive-summarizer) (1.0.2)
  Requirement already satisfied: catalogue<1.1.0,>=0.0.7 in /usr/local/lib/python3.6/dist-packages (from spacy->bert-extractive-summarizer) (1.0.0)
  Requirement already satisfied: srsly<1.1.0,>=1.0.2 in /usr/local/lib/python3.6/dist-packages (from spacy->bert-extractive-summarizer) (1.0.2)
  Requirement already satisfied: cymem<2.1.0,>=2.0.2 in /usr/local/lib/python3.6/dist-packages (from spacy->bert-extractive-summarizer) (2.0.3)
  Requirement already satisfied: blis<0.5.0,>=0.4.0 in /usr/local/lib/python3.6/dist-packages (from spacy->bert-extractive-summarizer) (0.4.1)
  Requirement already satisfied: six in /usr/local/lib/python3.6/dist-packages (from sacremoses->transformers->bert-extractive-summarizer) (1.12.0)
  Requirement already satisfied: click in /usr/local/lib/python3.6/dist-packages (from sacremoses->transformers->bert-extractive-summarizer) (7.1.2)
  Requirement already satisfied: urllib3!=1.25.0,!=1.25.1,<1.26,>=1.21.1 in /usr/local/lib/python3.6/dist-packages (from requests->transformers->bert-extractive-summarizer) (1.24
  Requirement already satisfied: idna<2.5 in /usr/local/lib/python3.6/dist-packages (from requests->transformers->bert-extractive-summarizer) (2.9)
  Requirement already satisfied: certifi>=2017.4.17 in /usr/local/lib/python3.6/dist-packages (from requests->transformers->bert-extractive-summarizer) (2020.6.20)
  Requirement already satisfied: chardet<4,>=3.0.2 in /usr/local/lib/python3.6/dist-packages (from requests->transformers->bert-extractive-summarizer) (3.0.4)
  Requirement already satisfied: pyparsing>=2.0.2 in /usr/local/lib/python3.6/dist-packages (from packaging->transformers->bert-extractive-summarizer) (2.4.7)
  Requirement already satisfied: importlib-metadata>=0.20; python_version < "3.8" in /usr/local/lib/python3.6/dist-packages (from catalogue<1.1.0,>=0.0.7->spacy->bert-extractive-
  Requirement already satisfied: zipp>=0.5 in /usr/local/lib/python3.6/dist-packages (from importlib-metadata>=0.20; python_version < "3.8"->catalogue<1.1.0,>=0.0.7->spacy->bert-
  Building wheels for collected packages: bert-extractive-summarizer
    Building wheel for bert-extractive-summarizer (setup.py) ... done
    Created wheel for bert-extractive-summarizer: filename=bert_extractive_summarizer-0.4.2-cp36-none-any.whl size=13711 sha256=4b41dbdaa5911361cb0050357829fdea8b27f33fe2ef089197
    Stored in directory: /root/.cache/pip/wheels/13/bc/30/654eb9e085717a56cba927c5a20b6cd01fb229b1ed2bf9b371
  Successfully built bert-extractive-summarizer
```

Figure 6-15. *Installation of packages*

2. Import the necessary packages required for this
    implementation, such as summarizer from the
    Summarizer library, using this command.

```
from summarizer import Summarizer
```

3. This library implements HuggingFace Pytorch
    transformers to run an extractive summarization. It
    works by generating embedding of sentences and
    then uses clustering algorithms such as a density-
    based algorithm, among others, to cluster sentences
    that are closest to centroids and form a highly dense
    region. Sentences from the highest density region
    will be taken to form the summary. Next, create an
    instance of Summarizer as shown here.

```
text_summarization_model = Summarizer()
```

4. Pass the document content as an argument to the Summarizer instance just created, as shown here.

```
return text_summarization_model(<Document Content>)
```

This will return a summary of document. Here is the complete Python code to perform document summarization using BERT.

```
from summarizer import Summarizer
def text_summary(text):
    model=Summarizer()
    return model(text)

if __name__=='__main__':
    text = "Machine learning (ML) is the study of computer
    algorithms that improve automatically through
    experience. It is seen as a subset of artificial
    intelligence. Machine learning algorithms build a
    mathematical model based on sample data, known as
    "training data", in order to make predictions or
    decisions without being explicitly programmed to do so.
    Machine learning algorithms are used in a wide variety
    of applications, such as email filtering and computer
    vision, where it is difficult or infeasible to develop
    conventional algorithms to perform the needed tasks."

    print(text_summary(text))
```

The text snippet in this example is from the Wikipedia article on machine learning.

Here is the resulting output:

```
Machine learning (ML) is the study of computer algorithms that
improve automatically through experience. It is seen as a
subset of artificial intelligence.
```

This output shows the summary of a document, and all of the sentences in summary are actual sentences from the document itself. A document can be of any length (e.g., 100 or 200 pages) and REST API won't be able to receive such an amount of data in a single API call. Therefore, as a best practice, a document summarization system should only be used as back-end application or system with a parent system such as a search engine, where every document returned as a part of the search result should have document summary as well.

The codebase for this exercise can be downloaded from GitHub at `https://github.com/bertbook/Python_code/tree/master/Chapter6/TextSummarization`.

Conclusion

This chapter covered the applicability of BERT in various NLP tasks such as sentiment analysis, text classification, entity recognition, and document summarization. We leveraged a BERT-based model to build NLP-based systems. In the next chapter, we will talk about the latest research happening in BERT.

CHAPTER 7

Future of BERT Models

The topics we have covered thus far deal with the architecture and application of the BERT model. The BERT model has not only affected the ML domain, but other fields like content marketing as well. Now let's discuss the development and future possibilities of BERT.

Future Capabilities

Transformer-based ML models like BERT have proven to be successful for state-of-the-art natural processing tasks. BERT, which is a large-scale model, remains one of the most popular language models that delivers state-of-the-art accuracy.

The BERT model has also been used by the Google search team to improve the query understanding capabilities of Google Search. As BERT is a bidirectional model, it is able to understand the context of a word by looking at the surrounding words. BERT is particularly helpful to capture the intent behind search queries.

Ever since its release, the BERT model has influenced the development of various models that are based on BERT. It has to be credited for the introduction of models that not only incorporate its name, but also its core architecture ideas. The variants of BERT are able to successfully beat

N. Sabharwal and A. Agrawal, *Hands-on Question Answering Systems with BERT*,
https://doi.org/10.1007/978-1-4842-6664-9_7

records across a wide array of NLP tasks like sentiment analysis, document classification, question answering, and more.

Here are a few of the models that are based on BERT.

- There are models that pertain to an application or domain-specific corpus. BioBERT is one such model that is trained on biomedical text. Other examples are SciBERT and Clinical BERT. Training on a domain-specific corpus has turned out to be useful and results in better performance when fine-tuning is done on downstream NLP tasks in contrast to fine-tuning BERT, which is trained on BookCorpus and Wikipedia.

- The ERNIE model incorporates knowledge into pretraining and uses a knowledge graph to mask entities and phrases. It is pretrained on a large corpus while taking the knowledge graph into consideration during input.

- The TransBERT model is used for a story ending prediction task that uses a three-stage unsupervised training approach. This is then followed by two supervised steps.

- For making medical recommendations, G-BERT basically combines the power of graph neural networks and BERT. This model is used for medical code recommendations and representations. Encoding of medical codes with hierarchical representations in G-BERT is done with the help of graph neural networks.

- In addition to pretrained models there are also fine-tuned models like DocBERT (document classification) and PatentBERT (patent classification). These models

are fine-tuned for specific tasks. These pretrained
BERT-based models can be fine-tuned with the help
of NLP tasks, POS, NER, and so on, to achieve better
results.

These models are representative of broad classes of BERT- based
models. They depict how the BERT model can further be used in different
domains with modifications in pretraining or fine-tuning. BERT hence
forms a base for the development of other models that are effective in a
wide variety of tasks.

One of the developments that relies on the BERT model is RoBERTa,
developed by Facebook, which has proven to be highly efficient on GLUE
benchmarking. RoBERTa uses the strategy of BERT to mask the text and the
machine learns to predict the hidden text. The training is done on a larger
number of mini-batches and learning rates, and the hyperparameters are
modified to achieve better results. These changes allowed the RoBERTa
model to prove its efficiency on MNLI, QNLI, RTE, STS-B, and RACE tasks,
and it also shows considerable improvement on the GLUE benchmark.

RoBERTa uses 160 GB of data for pretraining, which includes
unannotated NLP datasets and data scrapped from public news articles
called the CC-News dataset. These data, along with training of RoBERTa
on a 1024 V100 Tesla GPU, takes a day to complete. This results in better
performance of RoBERTa over other available models like BERT, XLNet,
Alice, and so on.

BERT is incorporated into Google Search, which results in precise and
accurate searches. This will affect the content strategy of many users. The
content now has to be more precise so that it can be rated better using
search engine optimization. The strategies to design the content have to be
improvised.

Abstractive Summarization

ML has come a long way in NLP, and one of these applications is in the field of summarization. The most common form of summarization is extractive summarization, which returns the most important sentences out of the content. The other type is abstractive summarization, which uses new sentences, keeping the important ideas or facts intact.

Content selection is integral to any summarization system. In recent approaches, the importance of separating content selection from summary generation is highly emphasized. There are many ongoing studies that attempt to extract content words and sentences that should be the part of summary and use them to guide the generation of an abstract summary.

A brief sentence can be formed by shortening or rewriting a lengthy text. Encoders and decoders are helpful in this context. Comprehensive summaries can be generated in a similar way, by selecting important sentences and dropping the inessential sentence elements, such as prepositional phrases. A summary can be generated through fusing multiple sentences. Selecting important sentences can be done via many approaches, but handling its large cardinality and identifying the sentence relationship to perform fusion has been a tough job. Previously it has been assumed that similar sentences can be fused together because they carry similar information to be processed.

Because abstractive summarization is difficult to perform, there is a lot of development in this area. BERT also has applications in abstractive summarization. The embeddings of multiple sentences can be generated using a BERT model. To perform this task, a [CLS] token can be inserted before the start of the first sentence. The output embeddings have to be processed through multiple layers, which enables the capture of important features. The BERTSUM model is one example.

Natural Language Generation

Natural language generation (NLG) is one of the more active research areas. It is a subgroup of NLP, along with NLU. The basic task of NLG is to convert some text tokens or data into natural language. The basic approach to achieve this is by predefining the templates for a specific domain and filling the empty slots using NLU techniques.

A more complex approach to this is using language modeling. Language modeling is used to model the natural language using the ways of writing, grammar, syntax, and so on, that are required to learn intrinsic features of the source language. We can then use this language in generating language content against some given input data or text.

The applications in terms of language understanding are not limited to NLP, but also extend to NLG. Open-AI's GPT-2 generates text based on the given words and is one of the state-of-the-art models in NLG. The BERT model tries to attain the same feature using HuggingFace transformers.

Recent developments show that the performance of BERT in the field of NLG is not an optial fit. The reason behind it is that the BERT model was trained on MLM rather than being trained autoregressively. Apart from using MLM, the variations such as shuffled input and random words make the BERT model more generalized. Even after all these variations, BERT lags behind GPT-2 because the BERT model is an encoder representation, whereas GPT-2 is a decoder stack, which helps it create context-rich representations.

Machine Translation

Translation is the idea of translating text from one language to another. Automatic or mechanical translation is probably one of the most challenging brain functions given the fluctuations in human language. Recently, pretraining techniques, such as ELMo, GPT and GPT-2, BERT, the cross-language model (XLM), XLNet, and RoBERTa have attracted a lot of attention in the ML and NLP communities.

A Neural Machine Translation (NMT) model usually consists of an encoder to map an input sequence to hidden representations, and a decoder to decode hidden representations and generate a sentence in the target language. BERT has achieved great success in NLU, and incorporating BERT with NMT for performance improvement might be a good research area.

NMT can be improved by fusing the BERT model and NMT, when BERT is drawn by the sensor and decoder using attention models. Research on open supervised NMT (including sentence-level and text-level translation), semisupervised NMT, and unsupervised NMT demonstrates the effectiveness of this approach.

To accurately predict translation quality, a model trained from scratch would theoretically require a large corpus of natural language source text, translations, and their human-labeled quality scores. Creating these datasets at a sufficient scale to train a neural network model is prohibitively expensive. Therefore, researchers have determined that they can transfer learnings from models trained on correctly translated parallel corpora to the task of identifying whether a translation is correct or not. It is far easier to obtain millions of correctly translated sentences to use to pretrain a model in areas where you don't need a quality score.

For future work, there are many interesting directions. First, we have to learn how to speed up the measurement process. Second, we can use such an algorithm in many applications, such as query in response. How to compress the BERT-fused model into a simplified version is another topic. There are other modern functions that include information about distillation to integrate pretrained models with NMT, which is a test method.

Conclusion

This chapter looked at the ongoing research in BERT and in state-of-the-art NLP tasks. With this we have come to the conclusion of this exciting journey into the world of NLP.

Index

A

Abstractive summarization, 167, 176
Activation function, 19
Advanced question answering systems, 98
ALBERT, 85, 87
Artificial intelligence (AI), 2
Artificial neural networks (ANN), 15, 16
Attention models
  global attention model, 38
  hard and soft attention model, 39
  local attention model, 38
  output vector of dense layer, 36
  requirement, 37
  self-attention model, 39
  working, 37, 38
Automatic routing of support tickets, 4

B

Backward propagation, 20, 21
Bag of words, 12, 13
Benchmarks, BERT model
  GLUE benchmark, 83

IMDB dataset, 84
  RACE, 85
  SQuAD dataset, 84
BERT, input representation, 102
BERT-based models
  ALBERT, 85, 87
  BERT$_{joint}$, 94, 95
  DistilBERT (*see* DistilBERT)
  RoBERTa, 88–90
  StructBERT (*see* StructBERT)
Bidirectional encoder–decoder architecture, 33
Bidirectional encoder representations from transformers (BERT)
  architecture, 60
  base model, 62
  embeddings, 67
  large model, 63
  masked language modeling, 67, 69
  models, 174
  next sentence prediction, 69–71
  text classification, 71–82
  text processing, 66, 67
Binary encodings, 41
BioBERT, 174

Printed in the United States
By Bookmasters